EVERYDAY ARITHMETIC

TEACH YOURSELF BOOKS

EVERYDAY ARITHMETIC

Peter and Kath Fearns

Hodder & Stoughton

A MEMBER OF THE HODDER HEADLINE GROUP

British Library Cataloguing in Publication Data

Fearns, Peter
   Teach yourself every arithmetic
   1. Arithmetic
   I. Title
   513
   ISBN 0 340 51155 9

First published 1990
Impression number   10 9 8 7
Year   1998 1997 1996

The 'Teach Yourself' name and logo are registered trade marks of
Hodder & Stoughton Ltd.

Typeset by MS Filmsetting Limited, Frome, Somerset.
Printed in Great Britain for Hodder & Stoughton Educational,
a division of Hodder Headline Plc, 338 Euston Road, London NW1 3BH
by Cox & Wyman Ltd, Reading, Berks.

Peter Fearns is a former head of department in Business Studies and is now Principal of Burton Upon Trent Technical College. He is the author of *Teach Yourself Business Studies*. Kath Fearns is a lecturer in Mathematics and collaborated with Peter in writing *Numeracy and Finance*.

Contents

Introduction

Introduction

The purpose of this book is to improve your arithmetic skills and your ability to deal with everyday calculations. The text is designed to help you to work on your own and at your own pace. You will find that the examples are taken from practical everyday experience. Not only will the book be useful for a wide range of activities such as shopping, gardening and DIY; it will also help you to gain confidence in basic mathematics.

The book is divided into two sections. Section One deals with the basic arithmetic skills such as measuring quantities and understanding percentages. Section Two covers common practical topics such as cooking, working and budgeting.

It is assumed, in Section One, that some readers' understanding of basic arithmetic is weak, and so the methods and procedures which are used in arithmetic are thoroughly explained. If you are not sure about the basic procedures in arithmetic, we would advise you to spend some time on this section, gaining confidence, before going on to Section Two. We have assumed that you will have access to a calculator, and, to help you, Chapter 3 explains how to use a calculator.

Although many everyday calculations are easy and straightforward, in some instances problems can seem quite complicated. For example, calculating the mix of proportions of garden chemicals or working out the actual cost of a loan can appear difficult. This book explains how to solve these and many other common problems and will help you to cope with everyday arithmetic.

SECTION ONE

Basic skills in arithmetic

1

Understanding numbers

Arithmetic is extremely important in helping to cope with problems we meet in everyday life. We use numbers in work, in shops, in the home, in travelling. Before dealing with problems in arithmetic we shall take a close look at the number system.

The symbols or digits **0, 1, 2, 3, 4, 5, 6, 7, 8, 9** which are used to represent numbers are familiar to all of us. By using a combination of the digits given above we can represent any value we want to.

The position of a digit in a number is important:

2387	2 represents	2000
1274	2 represents	200
6123	2 represents	20
7902	2 represents	2

Sometimes, when dealing with very large numbers, it is useful to place the digits into groups of three in order to make the numbers easier to read, for instance:

Two million can be written as 2 000 000

In order to work accurately with numbers it is useful to think of place value columns.

Example

	thousands	hundreds	tens	units
9082 means:	9	0	8	2

We also need to look at parts of numbers or numbers which are less than one. These numbers are called fractions. These are dealt with more fully in chapter 4. Decimal fractions also represent parts of whole numbers. Imagine one unit being split up further into ten parts: each part is therefore a tenth of a unit.

One of these parts would be written $\frac{1}{10}$

Two of these parts would be written $\frac{2}{10}$

Similarly if one unit is split up into a hundred parts then each part is a hundredth of a unit. One hundredth of a unit is written $\frac{1}{100}$, and three hundredths would be written $\frac{3}{100}$ and so on. These are called decimal fractions.

A decimal point (.) is used to separate whole numbers from the decimal fractions. The whole numbers are placed before the decimal point. The numbers after the decimal point represent tenths, hundredths, thousandths, etc.

Example
Place the following numbers in place value columns:

(a) 24   (b) 3.5   (c) 47.74   (d) 361.504

These are whole numbers			Decimal point	These are decimal fractions		
... *hundreds*	*tens*	*units*	.	*tenths*	*hundredths*	*thousandths* ...
H	T	U	.	t	h	th
	2	4				
		3	.	5		
	4	7	.	7	4	
3	6	1	.	5	0	4

Whole numbers can also be written in the form of decimals.

So, for example:

>35 can be written as 35.00
>4 can be written as   4.0000

Any number of noughts can be added after the decimal point at the end of a number. This does not change the value of the number.

>4.6
>4.60   all of these numbers
>4.600   have the same value.
>4.600000

Please note:
Any noughts after a decimal point but before other figures are important and affect the value of the whole number.

Example

>0.76
>0.076   all of these numbers
>0.000 76   have different values.
>0.000 007 6

2

Using numbers

There are four rules or operations which are the basic tools of arithmetic. These are:

1 Addition
2 Subtraction
3 Multiplication
4 Division

You should learn not only how to apply each of the rules, but when to apply each rule.

It is important to remember that each digit in a number has a particular value dependent on its position in the number; it is useful to associate numbers with columns of thousands, hundreds, tens, units, and so on (see page 2), whenever we add, subtract, multiply, or divide.

Addition

Addition allows us to find out how many there are altogether. The calculation is commonly known as 'adding up'. Other words which describe addition are:

'find the sum of'
'total'
'and'

The symbol used to show addition is the 'plus' sign (+).

Care must be taken when adding up numbers to ensure that they are correctly lined up so that units are added together, tens are added together and so on.

Example

The following numbers of tickets were sold for three charity football matches: **706, 115, 302**. How many were sold altogether?

Solution

Adding the numbers we get:

Th thousands	H hundreds	T tens	U units
	7	0	6
	1	1	5
	3	0	2
1	1	2	3

Add the units column first, then tens, then hundreds and so on.

It is not necessary to write down the place value columns each time. The important rule is to line the numbers up correctly.

Example

Find **4706 + 115 + 382**

Solution

$$
\begin{array}{r}
4706 \\
115 \\
382 \\
\hline
5203 \\
\hline
\end{array}
$$

The following example is included simply to show the importance of lining up the numbers correctly.

Example

Add **3.126, 7, 23.04**, and **2.1069**

Solution

$$
\begin{array}{r}
3.1260 \\
7.0000 \\
23.0400 \\
2.1069 \\
\hline
35.2729 \\
\hline
\end{array}
$$

You can put in noughts to make it easier to align the numbers

Now try the following calculations:

1. Add the following numbers: (a) 467, 52, 90, 16 (b) 7610 and 3192
   (c) 0.7, 21.29, 30.274 (d) 3, 4.12, 127.24, 1.376
   (e) 4.246, 20.678, 9.3
2. (a)   73.14 + 4.7 + 0.281 + 9.42   (b)   7.054 + 0.63 + 6.25 + 0.0067
   + 2.009
3. The cost of several items in a shopping basket is £3.49, £1.06, £1.98 and £4.27. Calculate the total amount.
4. At the beginning of a journey the tripometer reading of a car shows 12 903 miles. The owner of the car makes the following journeys: 27 miles and 112 miles. What will the new reading on the tripometer be?

Answers on page 167.

Subtraction

We subtract a number from another number when we want to take it away. Other words which mean the same as subtract are:

'find the difference between two numbers' 'take a number away from another number'	The symbol used for subtraction is the 'minus' sign $(-)$.

Example
A member of a sport's club is given the task of selling **20** raffle tickets. If she sells **14** how many tickets will she have left?

Solution
In order to find how many tickets are left we have to take 14 tickets away from the 20 tickets. i.e., we have to subtract 14 from 20:

$$20 - 14 = 6   (14 + 6 = 20)$$
There are 6 tickets left.

Care must be taken when subtracting numbers with several digits to ensure that they are correctly lined up.

Example
Find $18\,691 - 3270$

Solution

```
    1   8   6   9   1
-       3   2   7   0
  ─────────────────────
    1   5   4   2   1
```

Subtract units first, then tens, etc.

Example

A bank deposit account contained £2392. If £367 is withdrawn, how much money is left in the account?

Solution

We have to take 367 away from 2392, that is we must subtract 367 from 2392.

$$
\begin{array}{cccc}
2 & 3 & {}^{8}\cancel{9} & {}^{1}2 \\
- & 3 & 6 & 7 \\
\hline
2 & 0 & 2 & 5 \\
\end{array}
$$

Since we cannot take 7 from 2 reduce 9 tens to 8 tens and add the spare ten to the 2 to make 12. Now subtract 7 from 12 units, then 7 from 8 tens.

Alternative method:

$$
\begin{array}{cccc}
2 & 3 & 9 & {}^{1}2 \\
- & 3 & {}^{7}\cancel{8} & 7 \\
\hline
2 & 0 & 2 & 5 \\
\end{array}
$$

Since we cannot take 7 from 2, add 10 to the 2 to make 12, but to compensate add 10 to 367. This becomes 377. Then continue subtracting in the usual way.

You may be more familiar with the alternative method. You should use the method you prefer.

Example

Subtract 62.836 from 84.957

Solution

$$
\begin{array}{r}
84.957 \\
-62.836 \\
\hline
22.121 \\
\end{array}
$$

Example
Find $13.1 - 2.469$

Solution

$$\begin{array}{r} 13.100 \\ -2.469 \\ \hline 10.631 \\ \hline \end{array}$$ ⟵   add sufficient number of noughts.

Now try the following calculations:
1. (a) Find $1796 - 472$ (b) Calculate $2196 - 327$ (c) Subtract 295 from 1324.
2. Subtract 3.103 from 29.976
3. Take 2.104 from 3.605
4. $141.36 - 7.9$
5. $102.36 - 94.891$
6. 350 tickets are printed for a dance. If 293 tickets are sold how many tickets are left unsold?
7. The reading on an electricity meter at the beginning of the quarter is 52 624 units of electricity. At the end of the quarter the reading is 53 449, how many units of electricity have been used?

Answers on page 167.

Example
A listener wishes to record three radio programmes onto an audio-cassette. The first programme is **35** minutes long, the second is **28** minutes long and the third is **17** minutes long. How much space will be left on a **90** minute cassette after the three programmes have been recorded?

Solution
First of all work out how much of the tape will be used. This is found by adding together 35, 28, and 17.

$$35 + 28 + 17 = 80$$

so 80 minutes of the tape is used; by taking 80 minutes away from 90 minutes, that is subtracting, this will indicate how much tape is left.

$$90 - 80 = 10$$

There will be 10 minutes left unrecorded on the tape.

Example

A household keeps a small float of cash for sundry items. The float starts at £23.17, on the following day £6.12 is taken out, on the next day further sums of £10.98 and £2.01 are added to the float. Later, £7.29 is taken out to pay a further bill. How much money is there left in the float?

Solution

The problem given above can be written out as:

$$23.17 - 6.12 + 10.98 + 2.01 - 7.29$$

In problems which involve addition and subtraction together such as this, first add all the positive numbers, that is all of the numbers with a + sign or no sign in front:

```
 23.17
 10.98
  2.01
─────
 36.16
```
◄──── This gives the total money put into the 'float'.

Then ADD all of the negative numbers, that is all the numbers with a − sign in front:

```
 −6.12
 −7.29
─────
−13.41
```
◄──── This gives the total money taken out of the 'float'.

Now subtract the total of the negative numbers from the total of the positive numbers:

```
 36.16
−13.41
─────
 22.75
```
◄──── The final answer is £22.75, which is the money left in the 'float'.

Now try these calculations:

1. $3.46 - 7.1 + 15.362 - 2.69$
2. $9.86 + 27.361 - 41 - 3.95 + 8.365$
3. $1.63 - 2.97 - 9.617 + 3 + 7.4 + 5.91$

Answers on page 167.

Multiplication

The third rule is multiplication.

Suppose we have 3 fours, that is, 3 sets of four:

```
*   *   *   *
*   *   *   *
*   *   *   *
```

This is written as 3×4 and, by counting the stars, you can see that the total of 3 sets of 4 or $3 \times 4 = 12$. This process is called multiplication.

Example

Two rows of bedding plants are planted out with seven plants in each row. How many plants are there altogether?

Solution

2 sets of 7 are shown below:

```
*   *   *   *   *   *   *
*   *   *   *   *   *   *
```

Once again you can see that 2 sets of 7 or $2 \times 7 = 14$.

An alternative way of looking at multiplication is to add the numbers:

$$\text{Three 4s} = 4 + 4 + 4 = 3 \times 4 = 12$$
$$\text{Two 7s} = 7 + 7 = 2 \times 7 = 14$$

It would be tedious to add the different sets of numbers every time we want to multiply, so it is more convenient to learn the multiplication tables.

The multiplication table for numbers from 1 to 10 is given in the appendix, page 180.

Using the table we are able to multiply any 2 numbers between 1 and 10, for example 6×4:

$$\text{row } 6 \times \text{column } 4 = 24 \text{ and}$$
$$\text{row } 4 \times \text{column } 6 = 24$$

This is true for any pair of numbers. The order in which multiplication numbers are written is not important. We can multiply any number of figures together and the order in which it is done does not matter:

$$5 \times 6 \times 2 = 30 \times 2 = 60$$
$$6 \times 2 \times 5 = 12 \times 5 = 60$$

Now try these calculations:

1. Work out the following by (i) drawing a star diagram similar to the one on page 10, (ii) using the method of repeated addition and (iii) by using the multiplication table given in the appendix.
   (a) 3×6 (b) 4×9 (c) 5×8 (d) 1×7 (e) 9×6 (f) 8×9
2. Find the cost of 6 pens if each pen costs £2.00
3. You want to make 7 shelves, if each shelf requires 4 wall screws how many wall screws should you buy?
4. Take away sandwiches are sold in either brown or white bread, and each type of bread has a choice of 3 fillings, how many different types of sandwich are sold altogether?

Answers on page 167.

It is important to be able to multiply numbers by 10, 100, and so on. The examples given below show that this is straightforward.

Multiplying numbers by 10

The following examples show that when we multiply a number by 10 the place value of each digit changes. Each digit appears in the place value column ONE place to the LEFT.

Th	H	T	U	.	t	h		Th	H	T	U	.	t	h
			3				$\times 10 =$			3	0			
	6	7					$\times 10 =$		6	7	0			
7	2	0					$\times 10 =$	7	2	0	0			
		9		.	3		$\times 10 =$			9	3			
		0		.	7	8	$\times 10 =$			7		.	8	
	4	1	2	.	3	5	$\times 10 =$		4	1	2	3	.	5

Example
Find (a) 456×10 (b) 0.089×10 (c) 34.9521×10.

Solution

Following the pattern established above:

(a) $\quad\quad\quad\quad\quad$ 456 $\quad\times 10 = 4560$

(b) $\quad\quad\quad\quad\quad$ 0.089 $\times 10 = \quad\quad 0.89$

(c) $\quad\quad\quad\quad\quad$ 34.9521 $\times 10 = \quad 349.521$

Multiplication by 100

Consider the following:

$$1 \times 100 = \quad 100$$
$$2 \times 100 = \quad 200$$
$$38 \times 100 = 3800$$

Again we can see that the place value of the digits have changed. From the above we can see that multiplying by 100 has the effect of making the digits appear TWO places to the LEFT. For example:

$$67 \quad\quad\quad \times 100 = \quad 6\,700$$
$$720 \quad\quad\quad \times 100 = 72\,000$$
$$9.3 \quad\quad\quad \times 100 = \quad\quad 930$$
$$0.681 \quad\quad \times 100 = \quad\quad 68.1$$
$$412.3524 \times 100 = 41\,235.24$$

Other multiplication

To multiply by 1000, digits will appear THREE places to the LEFT:

$$214 \quad\quad\quad \times 1000 = 214\,000$$
$$0.17342 \times 1000 = \quad\quad 173.42$$
$$31.0079 \quad \times 1000 = \quad 31\,007.9$$

To multiply by 10 000, digits will appear FOUR places to the LEFT.

$$13 \quad\quad\quad\quad \times 10\,000 = \quad 130\,000$$
$$784.78066 \times 10\,000 = 7\,847\,806.6$$

Now try these calculations:

(a) 2.1×10 (b) 3.01×10 (c) 0.04×10 (d) 4.32×100 (e) 73.9×100
(f) 0.0061×100 (g) 6.53×1000 (h) 12.001×1000 (i) 0.03×1000
(j) $0.105\,34 \times 10\,000$ (k) $9.3 \times 10\,000$ (l) $0.0007 \times 100\,000$

Answers on page 167.

When you deal with complicated numbers it is much simpler to use a calculator. Make sure that you understand when to multiply numbers and that you are familiar with the section on calculators.

Division

The fourth rule in arithmetic is the division of numbers. We divide a number when we want to share it out equally.

Example
Suppose you want to share **£10** equally between **5** children

Solution
We have to divide the £10 by 5 and this can be shown in different ways:

$$10 \div 5 \quad or \quad \frac{10}{5} \quad or \quad 10/5 \quad or \quad 5\overline{)10}$$

There are 5 lots of £2 in £10 so that each child would receive £2.

You can see from this example that $5 \times 2 = 10$ and that $10/5 = 2$: division is the reverse of multiplication. You can always check your answer to a division problem by multiplying.

Example
There are **12** sandwiches on a plate to be shared between **3** people. How many sandwiches would each person get?

Solution
$$12/3 = 4$$

check $\qquad 3 \qquad \times \qquad 4 \qquad = \qquad 12$

no. of people × sandwiches = total sandwiches

Before you try the following exercises make sure that you are familiar with the multiplication tables on page 180.

Now try the following calculations:

1. (a) $14 \div 2$ (b) $9 \div 3$ (c) $28 \div 7$ (d) $30/5$ (e) $\dfrac{49}{7}$ (f) $36/4$

(g) $63 \div 9$　(h) $56 \div 7$　(i) $24/3$　(j) $\dfrac{60}{10}$

2. A curtain pole has 24 rings and you want to hang two curtains. How many rings should be used for each curtain?
3. A dish with 24 roast potatoes is to be shared between 6 people. How many potatoes should be served to each person?

Answers on page 167.

Dividing by 10

Division can be seen as the reverse of multiplication. When we divide a number by 10, the digits of the number appear ONE place to the RIGHT in the place value columns as shown in the following examples.

Th	H	T	U	.	t	h	th			Th	H	T	U	.	t	h	th
		4	0					$\div 10 =$					4				
	1	1	0					$\div 10 =$				1	1				
3	6	0	0					$\div 10 =$			3	6	0	.			
		6	7	.	3			$\div 10 =$					6	.	7	3	
			2	.	4	1		$\div 10 =$					0	.	2	4	1

Example
Find　(a) $10\,700 \div 10$　(b) $0.093 \div 10$

Solution
Using the pattern established above we have

(a)　　　　　　$10\,700 \quad \div 10 = 1070$
(b)　　　　　$0.093 \div 10 = \quad 0.0093$

Dividing by 100

When we divide a number by 100, the digits of the number appear TWO places to the RIGHT in the place value columns. Examine the following examples:

$$300 \quad \div 100 = 3$$
$$532.1 \quad \div 100 = 5.321$$
$$6.3 \quad \div 100 = 0.063$$
$$0.05 \div 100 = 0.0005$$

Using the principles in the above example it is fairly simple to divide even by 1000, 10 000, etc.

When we divide a number by 1000, the digits of the number appear THREE places to the RIGHT.

When we divide a number by 10 000, the digits of the number appear FOUR places to the RIGHT, etc.

Example

$$6\,809\,000/1000 = 6809$$
$$6\,809\,000/10\,000 = 680.9$$
$$45.02 \div 1000 = 0.04502$$
$$45.02 \div 10\,000 = 0.004502$$

Now try the following calculations:
1. (a) $3050 \div 10$ (b) $6740 \div 10$ (c) $5000 \div 10$
2. (a) $24\,800 \div 100$ (b) $74\,000 \div 100$ (c) $40\,600 \div 100$
3. (a) $235\,000 \div 1000$ (b) $7\,000\,000 \div 10\,000$ (c) $3\,675\,000 \div 100$
   (d) $4\,903\,000 \div 1000$ (e) $1\,000\,000 \div 100\,000$
   (f) $38\,000\,000 \div 1\,000\,000$
4. (a) $5.1 \div 10$ (b) $73.6 \div 10$ (c) $0.04 \div 10$ (d) $316 \div 100$
   (e) $87.49 \div 100$ (f) $0.601 \div 100$ (g) $938.2 \div 1000$
   (h) $0.01 \div 1000$ (i) $1.6 \div 10\,000$ (j) $39\,742.05 \div 10\,000$
   (k) $90.37 \div 100\,000$

Answers on page 167.

When you want to divide difficult numbers it is better to use your calculator.

Decimal places and accuracy

Sometimes numbers do not divide exactly.

Example

$$13/3$$

```
     4.333...
3 ⟌ 13.000
```

This division goes on indefinitely with 3 recurring each time. We say that the answer is 4.3 recurring (4.3˙).

Since an exact answer cannot be obtained you have to decide how accurate you wish the final answer to be, that is how many figures you require after the decimal point:

4.3   is correct to one decimal place, i.e. one figure after the decimal point. This can be written as 1 d.p.

4.33   is correct to two decimal places (2 d.p.).

4.333 is correct to three decimal places (3 d.p.).

In the example above we have **rounded off** the figures. We can round off any number to the degree of accuracy we want.

Consider the following numbers 2.8 and 2.1. If we want to round them off to whole numbers since 2.8 is closer to 3 than 2 then we should round it up to 3. But the number 2.1 is closer to 2 than 3 so we should round it down to 2.

Generally, since 2.5 is half-way between 2 and 3 any value greater than 2.5 should be rounded up to 3, and any value less than 2.5 should be rounded down to 2. Normally 2.5 is rounded up.

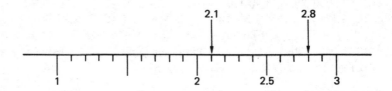

If we round off the following to the nearest whole numbers:

           12.3 becomes 12 (.3 is less than .5)
            3.7 becomes   4 (.7 is greater than .5)
            8.6 becomes   9 (.6 is greater than .5)
         17.4 becomes 17 (.4 is less than .5)
            6.5 becomes   7 (.5 is usually rounded up).

Now try these:

Write the following numbers to the nearest whole number:

(a) 12.6   (b) 3.2   (c) 10.5   (d) 1.32   (e) 12.65   (f) 0.93
(g) 100.09   (h) 8.497   (i) 0.01

Answers on page 168.

Rounding off to one decimal place

Rounding off to one decimal place means that we require one figure after the decimal point in the answer. To do this check the next number along; if it is 5 or more than 5 then round up, if it is less than 5 then round down.

Example
Give 3.61 to one decimal place.

Solution

3	.	6	1
		1 decimal place	this is **less than 5** so round down to 3.6

The answer is 3.6.

Example
Give 7.281 to one decimal place.

Solution

7	.	2	8	1
		1 decimal place	this is more than 5 so round up to 7.3	ignore this

The answer is 7.3.

Rounding off to two decimal places

This means that we require two figures after the decimal point.

Example
Write 6.7829 correct to two decimal places.

Solution

6	.	7	8	2	9
		2 decimal places		this is less than 5 so round down to 6.78	ignore this

The answer is 6.78.

Example
Write 2.3757 correct to two decimal places.

Solution

2	.	3 7	5	7
		2 decimal places	this is equal to 5 so round up to 2.38	ignore this

The answer is 2.38.

Examples

$$63.0289 \quad \text{is } 63.029 \quad \text{correct to 3 d.p.}$$
$$2.7649 \quad \text{is } 2.765 \quad \text{correct to 3 d.p.}$$
$$7.0085 \quad \text{is } 7.009 \quad \text{correct to 3 d.p.}$$
$$0.915\,04 \text{ is } 0.9150 \text{ correct to 4 d.p.}$$

Please note in the last example the last nought is included, this shows that it is accurate to 4 d.p.

$$10.369047 \text{ is } 10.36905 \text{ correct to 5 d.p.}$$

Now try these:
Write the following numbers correct to
(1) 1 d.p. and (2) 2 d.p.
(a) 16.423   (b) 0.069   (c) 3.0449   (d) 12.6014   (e) 9.057

Answers on page 168.

3

Using a calculator

Introduction

There are many different types of calculators available, from simple ones which deal only with basic operations $(+, -, \times, \div)$ to sophisticated programmable scientific and financial calculators. A simple calculator will allow you to do all the calculations in this book.

Calculators can be powered with ordinary batteries or solar cells. Calculators with solar cells require reasonable light and have the advantage that batteries do not have to be replaced. You should be careful not to damage the solar panel.

Calculators vary in the way they operate and so you should always read carefully the instruction booklet for your make of calculator. The following is a guide to the most common keys found on calculators.

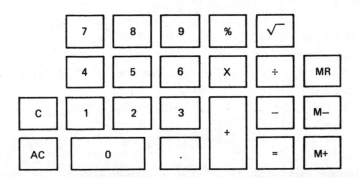

All calculators have number and $+$, $-$, $\times$, $\div$ keys. In addition your calculator will have some or all of the following keys:

ON/OFF	The OFF key is not always present on the solar powered calculator.
AC	This clears the calculator of any previous calculations.
C or CE	This clears the calculator of the last entry only, and allows you to continue with the calculation.
.	This is a decimal point.
√	This calculates the square root.
%	This calculates the percentage.
M+	This adds the calculation to the memory.
M−	This subtracts the calculation from the memory.
Min	This puts the calculation in the memory.
MR	This recalls the calculation from the memory.

Always remember to check that the contents of the memory are erased when you have finished the calculation.

Calculators are extremely quick and relatively simple to use. However, errors can easily occur by pressing the wrong buttons. Always try to work out an approximate value mentally for your calculation, then you can use this to check whether the answer on the calculator seems reasonable.

Example

Find 3.2×4

Solution

We know that $3 \times 4 = 12$. Since 3.2 is close in value to 3 then 3.2×4 should give an answer close to 12.

The calculation is done by pressing the buttons in the following sequence:

$$\boxed{3} \boxed{.} \boxed{2} \boxed{\times} \boxed{4} \boxed{=}$$

The answer on the display is $\boxed{12.8}$

Example
Find 24.381 ÷ 5.67

Solution
First make a rough calculation:
  Rounding off the numbers above to whole numbers we get

$$24 \div 6$$

The answer to this is 4, so we expect 24.381 ÷ 5.67 to be approximately in the region of 4 as opposed to 40 or 0.4.
  Using the calculator:

$$\boxed{2}\ \boxed{4}\ \boxed{.}\ \boxed{3}\ \boxed{8}\ \boxed{1}\ \boxed{\div}\ \boxed{5}\ \boxed{.}\ \boxed{6}\ \boxed{7}\ \boxed{=}$$

The answer on the display is $\boxed{4.3}$

Example
A packet of biscuits contains **23** biscuits. If we have **4** packets, how many biscuits are there altogether.

Solution
To find the total number we need to calculate four 23s which is 4 × 23.
  This is done by pressing the calculator buttons in the following sequence:

$$\boxed{4}\ \boxed{\times}\ \boxed{2}\ \boxed{3}\ \boxed{=}$$

The answer on the display is $\boxed{92}$

Example
Calculate 3210/15

Solution
The buttons should be pressed in the following sequence:

$$\boxed{3}\ \boxed{2}\ \boxed{1}\ \boxed{0}\ \boxed{\div}\ \boxed{1}\ \boxed{5}\ \boxed{=}$$

The answer on the display is $\boxed{214}$

Example
Find 2.43 × 10.126

Solution

This is done by pressing the buttons in the following sequence:

$$\boxed{2}\ \boxed{.}\ \boxed{4}\ \boxed{3}\ \boxed{\times}\ \boxed{1}\ \boxed{0}\ \boxed{.}\ \boxed{1}\ \boxed{2}\ \boxed{6}\ \boxed{=}$$

The answer on the display is $\boxed{24.60618}$

Example

Find $24.49 + 5.72 - 3.641$

$$\boxed{2}\ \boxed{4}\ \boxed{.}\ \boxed{4}\ \boxed{9}\ \boxed{+}\ \boxed{5}\ \boxed{.}\ \boxed{7}\ \boxed{2}\ \boxed{-}\ \boxed{3}\ \boxed{.}\ \boxed{6}\ \boxed{4}\ \boxed{1}\ \boxed{=}$$

The answer on the display is $\boxed{26.569}$

Now try these:

Using your calculator find:

(a) $2.36 + 4.561 + 23.4$ (b) $9.06 + 1.37 - 3.612$ (c) $34.1 - 26.095$
(d) 241×39.97 (e) $3.21 \times 2.78 \times 3.1$ (f) $14.99/8.14$

Answers on page 168.

Order of operations

You may be faced with calculations involving several operations. The order in which the calculation is done is important. If the numbers are keyed into the calculator simply in the order they are written down this may result in an incorrect answer. The order in which a calculation should be done is as follows:

(1) Brackets (if there are any)
then (2) multiplication and division
then (3) addition and subtraction.

This rule is sometimes called the BODMAS rule, that is, **B**rackets, **O**f, **D**ivision, **M**ultiplication, **A**ddition, **S**ubtraction.

Some calculators use the BODMAS rule automatically, but many do not. If you calculator does not use the BODMAS rule then you must make sure that you key the numbers into the calculator in the correct sequence. You should first check whether your calculator uses the BODMAS rule using the example given below.

Example

Find $3 + 4 \times 6$

Solution
Try the following sequence:

$$\boxed{3} \boxplus \boxed{4} \boxtimes \boxed{6} \boxminus$$

The correct answer is 27. If your calculator does not give this value it has not used the BODMAS rule. You should take care to key in the numbers in the correct sequence as follows:

$\boxed{4} \boxtimes \boxed{6} \boxplus \boxed{3} \boxminus$ Multiplication first, then Addition

The answer is $\boxed{27}$

Example
Find $(15 + 3)/6$

Solution
$\boxed{1}\boxed{5} \boxplus \boxed{3} \boxminus \boxdiv \boxed{6} \boxminus$ Brackets first, then Division

The answer is $\boxed{3}$

Example
Find $15 \div 3 + 4.2 \times 12 - 10$

Solution
(i) Do multiplication and division first.

$$\boxed{1}\boxed{5} \boxdiv \boxed{3} \boxminus$$

The answer is $\boxed{5}$. Write this answer down or place it in the memory
$\boxed{\text{Min}}$

$$\boxed{4} \boxed{.} \boxed{2} \boxtimes \boxed{1}\boxed{2} \boxminus$$

The answer is $\boxed{50.4}$

(ii) Now work out any addition or subtraction.

$$50.4 \boxplus \boxed{\text{MR}} \boxminus \boxed{1}\boxed{0} \boxminus$$

This is already in the display	Recall **5** from memory.

The final answer is $\boxed{45.4}$

Now try these calculations:
(a) $17.1 - (4.3 + 2.1)$ (b) $20.4 + 2 \times 13$ (c) $3 \times 4.1 + 5.6 \div 2$
(d) $2 \times (4 + 1.6) - 3 \times 6 + 5$ (e) $8 \times 3 + 4 - (10 + 12 \div 3)$

Answers on page 168.

Dealing with money and quantities

You often need the decimal point key $\boxed{.}$ to do questions which deal with money or other quantities.

Some calculators leave out zeros after the decimal point.

Example
Find £2.55 + £3.45

Solution

$$\boxed{2}\ \boxed{.}\ \boxed{5}\ \boxed{5}\ \boxed{+}\ \boxed{3}\ \boxed{.}\ \boxed{4}\ \boxed{5}\ \boxed{=}$$

The answer on the display is given as $\boxed{6}$. The answer should be written £6.00.

Example
Find £2.14 + £5.56

Solution

$$\boxed{2}\ \boxed{.}\ \boxed{1}\ \boxed{4}\ \boxed{+}\ \boxed{5}\ \boxed{.}\ \boxed{5}\ \boxed{6}\ \boxed{=}$$

The answer given as $\boxed{7.7}$. This should be written £7.70.

You should take care when working in different units such as £ and p, or kg and g. When you use a calculator it is assumed that all the amounts keyed in are in the same units.

Example
Find £2.21 + 45p

Solution
If you key these numbers in as they are the answer given in the display will be $\boxed{47.21}$. This is obviously wrong because the calculator added £2.21 + £45. You should first change 45p to £0.45
    £2.21 + £0.45 = £2.66 which is correct.

Example
Find 86p − 15p + 37p

Solution
The answer on the display is ⌗108⌗. This should be written as £1.08.

Example
Add **3** litres, **500** ml, and **1.6** litres

Solution
You should change the units either to litres or millilitres.
    500 ml can be written as 0.5 litres. We now have

$$\boxed{3}\ \boxed{+}\ \boxed{.}\ \boxed{5}\ \boxed{+}\ \boxed{1}\ \boxed{.}\ \boxed{6}\ \boxed{=}$$

The answer is 5.1 litres.
See Chapter 5 on measuring qualities.

Now try these calculations:
(a) £2.61 + 75p   (b) £7.53 + £1.67   (c) 23p + 46p + 75p
(d) £1.26 − 37p   (e) £9.56 + 4p + 80p   (f) 450 ml + 2l + 1.6l
(g) 300 g + 2 kg − 750 g

Answers on page 168.

4

Fractions

Fractions, like decimals, represent parts of whole numbers or numbers which are less than 1. Going metric has meant that there is a greater emphasis on decimals, nevertheless, in some everyday situations it is useful to be able to manipulate simple fractions.

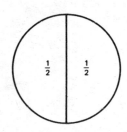

1 whole cake can be cut up into smaller parts or fractions of a cake. If it is cut into two pieces we have two halves. One half is written as $\frac{1}{2}$. This means 1 whole divided into two parts.

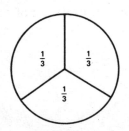

$\frac{1}{3}$ means that 1 whole is divided into 3 parts. $\frac{1}{3}$ represents one of the parts.

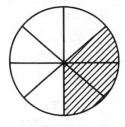

$\frac{1}{8}$ means that 1 whole is divided into 8 parts. $\frac{1}{8}$ represents one of the parts. The shaded area in the diagram shows 3 of these, parts this is $\frac{3}{8}$.

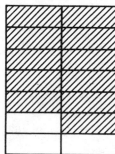

This shaded area represents $\frac{11}{14}$.

Now try the following:
What fraction of the following areas are shaded?

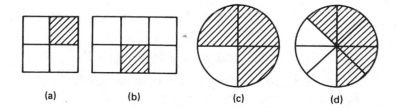

(a)   (b)   (c)   (d)

Answers on page 168.

Whole numbers can be written as fractions with a denominator of 1. For example $4 = \frac{4}{1}$, $7 = \frac{7}{1}$ and so on.

The number written on top of the fraction is called the **numerator** and the bottom number is called the **denominator**. For example

$$\frac{7}{16} \quad \begin{array}{l} \text{the numerator is 7} \\ \text{the denominator is 16} \end{array}$$

You can multiply or divide the numerator and denominator of a fraction by the same number without changing the value of the fraction.

For example

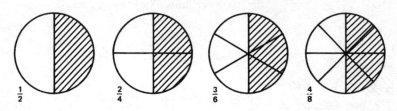

$\frac{1}{2}$ $\frac{2}{4}$ $\frac{3}{6}$ $\frac{4}{8}$

All the shaded areas are equal.

So that $\frac{1}{2} = \frac{2}{4} = \frac{3}{6} = \frac{4}{8}$

$\frac{1}{2}$ can be obtained directly by dividing the numerator and the denominator of:

$$\frac{2}{4} \text{ by } 2$$
$$\frac{3}{6} \text{ by } 3$$
$$\frac{4}{8} \text{ by } 4$$

Consider the following:

$\frac{5}{10}$ dividing through by 5 gives $\frac{1}{2}$, so $\frac{5}{10} = \frac{1}{2}$

$\frac{2}{6}$ dividing through by 2 gives $\frac{1}{3}$, so $\frac{2}{6} = \frac{1}{3}$

$\frac{12}{16}$ dividing through by 4 gives $\frac{3}{4}$, so $\frac{12}{16} = \frac{3}{4}$

Dividing the numerator and denominator by the same number in this way is sometimes called 'cancelling down'.

Similarly if we multiply the numerator and the denominator of a fraction by the same number the value of the fraction stays the same. For example:

$\frac{1}{2}$ multiplying through by 4 gives $\frac{4}{8}$, so $\frac{1}{2} = \frac{4}{8}$

$\frac{2}{3}$ multiplying through by 5 gives $\frac{10}{15}$, so $\frac{2}{3} = \frac{10}{15}$

Now complete the following:

(a) $\dfrac{1}{4} = \dfrac{3}{8} = \dfrac{}{} = \dfrac{}{16}$

(b) $\dfrac{3}{5}=\dfrac{}{10}=\dfrac{}{30}=\dfrac{15}{}$

(c) $\dfrac{2}{14}=\dfrac{}{7}=\dfrac{3}{}=\dfrac{}{56}$

(d) $\dfrac{10}{12}=\dfrac{}{6}=\dfrac{15}{}=\dfrac{}{48}$

Answers on page 168.

Proper and Improper fractions

If a fraction has a numerator which is smaller than the denominator, it is known as a **proper fraction**.
  For example $\frac{6}{7}$, $\frac{2}{5}$, $\frac{5}{9}$, $\frac{112}{151}$, are all proper fractions.

If a fraction has a numerator which is larger than the denominator then it is known as an **improper fraction**.
  For example $\frac{11}{6}$, $\frac{24}{16}$, $\frac{5}{2}$ are all improper fractions.
The meaning of an improper fraction is shown in the diagram below. $\frac{11}{6}$ means 11 'lots' of $\frac{1}{6}$.

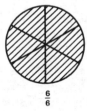
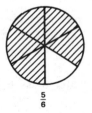

$\frac{6}{6}$ $\qquad$ $\frac{5}{6}$

  From the diagram you can see that $\frac{11}{6}$ is the same as 1 whole plus $\frac{5}{6}$. This is written as $1\frac{5}{6}$.
  Since $1\frac{5}{6}$ consists of a whole number and a fraction it is called a **mixed number**.
  For example:

$$12\tfrac{3}{4},\ 2\tfrac{7}{8},\ 24\tfrac{1}{7}$$

are all mixed numbers.

It is useful to be able to change improper fractions to mixed numbers, and mixed numbers to improper fractions.

Changing improper fractions to mixed numbers

To change an improper fraction to a mixed number you should first divide the denominator into the numerator, this gives the whole number, and check the remainder. This remainder is then written as a fraction of the denominator.

Example
Write the following fractions as mixed numbers:

(a) $\frac{13}{5}$ (b) $\frac{26}{4}$

Solution

(a) $\frac{13}{5}$ If we divide 5 into 13 we get 2 and a remainder of 3.
So $\frac{13}{5} = 2\frac{3}{5}$

(b) $\frac{26}{4}$ If we divide 4 into 26 we get 6 and the remainder is 2.
So $\frac{26}{4} = 6\frac{2}{4} = 6\frac{1}{2}$

Changing mixed numbers to improper fractions

To change from mixed numbers to improper fractions, reverse the operations above as the following examples show.

Example
Change the following mixed numbers to improper fractions:
(a) $2\frac{3}{5}$ (b) $5\frac{7}{9}$

Solution

(a)
$$2\frac{3}{5} = 2 + \frac{3}{5} \quad \text{But } 2 = \frac{10}{5}$$
$$\text{so } 2 + \frac{3}{5} = \frac{10}{5} + \frac{3}{5} = \frac{13}{5}$$

The required fraction is $\frac{13}{5}$

(b)
$$5\frac{7}{9} = 5 + \frac{7}{9} \quad \text{But } 5 = \frac{45}{9}$$
$$\text{so } 5 + \frac{7}{9} = \frac{45}{9} + \frac{7}{9} = \frac{52}{9}$$

The required fraction is $\frac{52}{9}$

Now try the following:
1. Change the following improper fractions to mixed numbers:

(a) $\frac{5}{2}$ (b) $\frac{12}{7}$ (c) $\frac{4}{3}$ (d) $\frac{36}{7}$

2. Change the following mixed numbers to improper fractions:

(a) $5\frac{2}{7}$ (b) $1\frac{5}{8}$ (c) $9\frac{3}{4}$ (d) $1\frac{13}{17}$

Answers on page 168.

Changing decimals to fractions

Example
From chapter one we know that:

$14.2 = 14\frac{2}{10} = 14\frac{1}{5}$

$23.7 = 23\frac{7}{10}$

$1069.1 = 1069\frac{1}{10}$

$37.54 = 37\frac{54}{100} = 37\frac{27}{50}$

$419.06 = 419\frac{6}{100} = 419\frac{3}{50}$

$1.99 = 1\frac{99}{100}$

$54.577 = 54\frac{577}{1000}$

$16.103 = 16\frac{103}{1000}$

$146.008 = 146\frac{8}{1000} = 146\frac{1}{125}$

Now try these:
Change the following decimals to fractions. Do not forget to cancel down where possible.

(a) 29.4   (b) 261.3   (c) 0.9   (d) 3.61   (e) 40.75   (f) 9.04
(g) 12.362   (h) 7.702   (i) 38.117   (j) 0.0018

Answers on page 169.

Changing fractions to decimals

This is straightforward when the fractions are in tenths, hundredths, and so on. For example:

$\frac{1}{10}=0.1$ $\frac{7}{10}=0.7$ $3\frac{9}{10}=3.9$ $6\frac{54}{100}=6.54$ $\frac{3}{100}=0.03$

$12\frac{13}{1000}=12.013$

Now try these:

Change the following fractions to decimals:

(a) $1\frac{3}{10}$ (b) $21\frac{7}{10}$ (c) $6\frac{39}{100}$ (d) $13\frac{7}{100}$ (e) $\frac{53}{100}$ (f) $5\frac{21}{1000}$

(g) $8\frac{984}{1000}$ (h) $\frac{1284}{10000}$

Answers on page 169.

In order to change any fractions to decimals write the numerator (the number on the top) as a decimal and then divide it by the denominator (the number on the bottom).

Example

Change $\frac{4}{5}$ to a decimal.

Solution

Write 4 as a decimal (use as many noughts as you need to divide through completely). Then divide by 5.

$$4.0 \div 5 = 0.8$$

So $\frac{4}{5}=0.8$

Example

Change $12\frac{7}{8}$ to a decimal.

Solution

First convert $\frac{7}{8}$ to a decimal. Write 7 as a decimal and then divide by 8:

$$7.000 \div 8 = 0.875$$

Now add the whole number 12
$12\frac{7}{8} = 12.875$

Now try these:
Change the following fractions to decimals:

(a) $\frac{3}{4}$ (b) $\frac{2}{5}$ (c) $\frac{1}{2}$ (d) $\frac{4}{15}$ (e) $\frac{3}{8}$ (f) $\frac{5}{16}$ (g) $4\frac{1}{4}$ (h) $7\frac{1}{5}$

(i) $13\frac{7}{20}$ (j) $9\frac{5}{8}$ (k) $3\frac{23}{25}$

Answers on page 169.

Addition and subtraction of fractions

Fractions can be added or subtracted directly provided they have the same denominators.
For example:

(a) $\frac{4}{5} + \frac{3}{5} = \frac{7}{5} = 1\frac{2}{5}$

(b) $\frac{2}{15} + \frac{8}{15} = \frac{10}{15} = \frac{2}{3}$

(c) $1\frac{1}{8} + \frac{5}{8} + \frac{3}{8} = \frac{9}{8} + \frac{5}{8} + \frac{3}{8} = \frac{17}{8} = 2\frac{1}{8}$ Note: we only add the numerators

Now try the following calculations:

(a) $\frac{3}{8} + \frac{1}{8} + \frac{5}{8}$ (b) $\frac{4}{11} + \frac{6}{11} + \frac{9}{11} + \frac{1}{11}$ (c) $2\frac{1}{5} + \frac{3}{5} + \frac{4}{5}$ (d) $\frac{5}{9} + 1\frac{4}{9}$

Answers on page 169.

For example:

(a) $\frac{5}{8} + \frac{3}{8} = \frac{2}{8} = \frac{1}{4}$ Note: we only subtract

(b) $1\frac{5}{8} - \frac{7}{8} = \frac{13}{8} - \frac{7}{8} = \frac{6}{8} = \frac{3}{4}$ the numerators

Now try the following calculations:

(a) $\dfrac{4}{9} - \dfrac{1}{9}$ (b) $\dfrac{4}{7} - \dfrac{2}{7}$ (c) $\dfrac{11}{15} - \dfrac{7}{15}$ (d) $1\dfrac{1}{6} - \dfrac{5}{6}$

Answers on page 169.

If fractions have different denominators you should first change the denominators of one or both fractions to make them equal. The fractions can then be added or subtracted together.

Example

Find $\dfrac{2}{5} + \dfrac{3}{4}$

Solution

We must first find the smallest number that both 5 and 4 will divide into. This number is 20. Now change both denominators to 20. Then add the fractions in the usual way.

$$\tfrac{2}{5} = \tfrac{8}{20} \quad \tfrac{3}{4} = \tfrac{15}{20}$$
$$\tfrac{2}{5} + \tfrac{3}{4} = \tfrac{8}{20} + \tfrac{15}{20} = \tfrac{23}{20}$$
$$\tfrac{23}{20} = 1\tfrac{3}{20}$$

Example

A DIY enthusiast has to fix a piece of wood to a wall. The thickness of the wood is $1\frac{1}{2}$ ins and it is recommended that the screws which fix the wood should be fixed at a depth of $\frac{3}{4}$ ins into the wall. What length of screw should be used?

Solution

Before we add any mixed numbers they should be changed to improper fractions.

$$1\tfrac{1}{2} + \tfrac{3}{4} = \tfrac{3}{2} + \tfrac{3}{4}$$
$$\tfrac{3}{2} + \tfrac{3}{4} = \tfrac{6}{4} + \tfrac{3}{4} = \tfrac{9}{4}$$
$$\tfrac{9}{4} = 2\tfrac{1}{4}$$

The screws should be $2\frac{1}{4}$ inches long.

Example

Find $\frac{3}{8}+\frac{5}{6}-\frac{3}{4}$

Solution

First find the smallest number that 8, 6, and 4 will divide into. This number is 24. Now change all denominators to 24, then add the fractions in the following way.

$$\frac{3}{8}=\frac{9}{24}; \quad \frac{5}{6}=\frac{20}{24}; \quad \frac{3}{4}=\frac{18}{24}$$

So $\frac{3}{8}+\frac{5}{6}-\frac{3}{4}=\frac{9}{24}+\frac{20}{24}-\frac{18}{24}=\frac{11}{24}$

Now try the following calculations:

(a) $\frac{1}{6}+\frac{3}{4}$ (b) $\frac{2}{5}+\frac{5}{6}$ (c) $\frac{3}{7}+\frac{1}{4}+\frac{1}{2}$ (d) $1\frac{1}{4}+2\frac{1}{3}$ (e) $\frac{7}{8}+3\frac{1}{5}$

(f) $1+4\frac{1}{6}+2\frac{3}{8}$ (g) $\frac{3}{5}-\frac{4}{15}$ (h) $1\frac{3}{4}-\frac{2}{7}$ (i) $4\frac{5}{6}-1\frac{3}{7}$

Answers on page 169.

Multiplication of fractions

Consider the shaded area shown.
The area represents $\frac{2}{3}$

but the area is also equal to 2 'lots' of $\frac{1}{3}$,
i.e. $2\times\frac{1}{3}$

so $2\times\frac{1}{3}=\frac{2}{1}\times\frac{1}{3}=\frac{2}{3}$

This gives us a rule for multiplying fractions: first multiply the numbers in the numerator together (2×1), then multiply the numbers in the denominator together (1×3).

Similarly

Consider $\frac{1}{2}$ of $\frac{1}{3}$. This is equal to $\frac{1}{6}$, i.e. $\frac{1}{2} \times \frac{1}{3} = \frac{1}{6}$

Any mixed numbers should first be changed to improper fractions and then multiplied in the usual way.

Examples
Find $1\frac{2}{3} \times 2\frac{2}{5}$

Solution
$1\frac{2}{3} \times 2\frac{2}{5} = \frac{5}{3} \times \frac{12}{5} = \frac{60}{15} = 4$

Now try the following calculations:
(a) $\frac{2}{3} \times 2\frac{1}{4}$ (b) $3\frac{5}{8} \times 1\frac{1}{3}$ (c) $\frac{2}{15} \times 3\frac{3}{4} \times 1\frac{1}{4}$

Answers on page 169.

Dividing fractions

Example
A length of fabric is $4\frac{1}{2}$ yds long. It is to be cut into 3 equal lengths. How long would each length be?

Solution

This can be written in two ways

$$\text{cut length} = 4\tfrac{1}{2} \div 3$$

or

We know that each cut length will be $\tfrac{1}{3}$ of the original length.

So cut length $= \tfrac{1}{3} \times 4\tfrac{1}{2}$

It follows that

$$4\tfrac{1}{2} \div 3 = 4\tfrac{1}{2} \times \tfrac{1}{3}$$
$$4\tfrac{1}{2} \times \tfrac{1}{3} = \tfrac{9}{2} \times \tfrac{1}{3} = \tfrac{9}{6} = \tfrac{3}{2}$$
$$\tfrac{3}{2} = 1\tfrac{1}{2}$$

Each cut length is $1\tfrac{1}{2}$ yds.

The simple rule for dividing fractions is: change division sign ($\div$) to multiplication sign ($\times$) and invert the fraction you wish to divide by.

Example

Find $\tfrac{5}{7} \div \tfrac{3}{4}$

Solution

$\tfrac{5}{7} \div \tfrac{3}{4} = \tfrac{5}{7} \times \tfrac{4}{3} = \tfrac{20}{21}$

Now try the following calculations.

(a) $1\dfrac{5}{7} \div \dfrac{3}{14}$ (b) $5\dfrac{1}{5} \div 1\dfrac{3}{10}$ (c) $2\dfrac{5}{9} \div 7\dfrac{2}{3}$

(d) $4\dfrac{5}{6} \div 1\dfrac{5}{18}$ (e) $6\dfrac{3}{8} \div 3$ (f) $5 \div 3\dfrac{1}{3}$

Answers on page 169.

5

Measuring quantities

When we measure quantities a choice of either imperial units or metric units is available. Metric units are considerably easier to use and their use is becoming increasingly common. There are many different units within the metric system; the most important ones are dealt with in this book. Since imperial units are still occasionally in use, some reference is made to these.

Mass

The standard **metric system** unit for weighing is a gram. For instance, the amounts used in a recipe would be measured in grams. Small amounts are weighed in milligrams; medicines are often weighed in milligrams. For heavier weights such as vegetables or fruit then these are weighed in kilograms. Very heavy loads are weighed in metric tonnes.

There are:

        1000 milligrams in 1 gram
        1000 grams in   1 kilogram
        1000 kilograms in 1 metric tonne

        Milligrams are shortened to mg
        Grams   are shortened to g
        Kilograms are shortened to kg

Example
How many milligrams are there in 0.34 g?

Solution

$$1\,g = 1000\,mg$$

so $0.34\,g = 0.34 \times 1000\,mg = 340\,mg$.

Example

How many kilograms are there in (a) 6000 g   (b) 835 g?

Solution

(a)
$$1000\,g = 1\,kg$$
$$\text{so } 1\,g = \tfrac{1}{1000}\,kg$$
$$6000\,g = 6000 \times \tfrac{1}{1000}\,kg = 6\,kg$$

(b) Similarly $835\,g = 835 \times \tfrac{1}{1000} = 0..835\,kg$

In the **imperial system**, ounces (oz), pounds (lb), stones, hundred-weights (cwt) and tons are used to weigh quantities. A full table of imperial weights and measures is given in the appendix, page 180.

Example

A man weighs 12 stones 10 lb. Express this in pounds.

Solution

From the table we see that 1 stone = 14 lb

$$\text{so } 12 \text{ stones} = 12 \times 14\,lb = 168\,lb$$

so 12 stones 10 lb = 168 lb + 10 lb = 178 lb.

Example

Express 4 lb 5 oz in pounds.

Solution

From the table we see that 16 oz = 1 lb

$$\text{so } 1\,oz = \tfrac{1}{16}\,lb$$
$$\text{so } 5\,oz = \tfrac{5}{16}\,lb = 0.3125\,lb$$

so 4 lb 5 oz = 4.3125 lb.

Capacity or volume

Capacity is a way of measuring a space taken up by liquids. The standard unit for measuring capacity in the **metric system** is a litre. Smaller quantities are measured in millilitres or centilitres.

There are:

1000 millilitres in 1 litre
100 centilitres in 1 litre
10 millilitres in 1 centilitre

Millilitres	are shortened to ml
Centilitres	are shortened to cl
Litres	are shortened to l

Example

Change 0.75 litres to centilitres.

Solution

$$1 l = 100 cl$$

so $0.75 l = 0.75 \times 100 cl = 75 cl$.

The table for imperial units and the conversion of imperial to metric are given in the appendix, page 180.

Example

A standard bottle of wine measures 75 centilitres. Convert this to pints.

Solutions

From the table we see that 4.546 litres = 1 gallon
since 1 gallon = 8 pints then 4.546 litres = 8 pints
so 1 litre = 8/4.546 pints = 1.76 pints
so 1 centilitres = 1.76 pints

so 75 centilitres = $75 \times 1.76/100$ pints = 1.32 pints.

Lengths and distances

In a **metric system** the standard unit for measuring length is the metre. Given below is a table of the most common units which are associated with the metre. The abbreviation for each unit is given in brackets.

$$10 \text{ millimetres (mm)} = 1 \text{ centimetre (cm)}$$
$$1000 \text{ millimetres (mm)} = 1 \text{ metre (m)}$$
$$100 \text{ centimetres (cm)} = 1 \text{ metre (m)}$$
$$1000 \text{ metres} \quad \text{(m)} \quad = 1 \text{ kilometre (km)}$$

Small lengths such as cross sections of timber are measured in millimetres or centimetres, large lengths such as the dimensions of a room are measured in metres, and distances are measured in kilometres.

Example
A length of timber is 3 m long. How many shelves of length 82 cm can be cut from this piece of wood?

Solution
$$1 \text{ m} = 100 \text{ cm and so } 3 \text{ m} = 300 \text{ cm}$$

We need to calculate how many lengths of 82 cm there are in 300 cm.

so the number of shelves is $300/82 = 3.66$

3 shelves could be cut from the 3 m length.

In the imperial system inches (ins or ″), feet (ft or ′), yards (yd) and miles are used to measure lengths and distances. A full table of imperial weights and measures is given in the appendix, page 180.

Example
The speed restriction on a road in France is given as 50 km per hour. What is the equivalent speed in miles per hour?

Solution
We have to find how many miles are equivalent to 50 km. From the table in the appendix we can see that

$$1 \text{ km} = 0.6214 \text{ miles}$$
$$\text{so } 50 \text{ km} = 50 \times 0.6214 \text{ miles} = 31.07 \text{ miles}$$

50 km per hour is approximately 31 miles per hour.

The importance of common units

It is important in calculations to make sure that all measurements are quoted in the same units before they are added, subtracted, multiplied or divided.

Example
Add the following amounts:

$$2\,kg, 250\,g, 3.2\,kg, 460\,g$$

Solution
Some of these quantities are measured in grams and others in kilograms. These cannot be added directly together; they should all be changed to either grams or kilograms whichever is appropriate. Changing grams to kilograms: $250\,g = 0.25\,kg$ and $460\,g = 0.46\,kg$. The total weight is $2\,kg + 0.25\,kg + 3.2\,kg + 0.46\,kg = 5.91\,kg$.

Now try these:
1. Change 2.3 litres to millilitres.
2. How many 5 ml spoonfuls are there in 1 cl?
3. How many kilograms are there in 4600 grams?
4. Add 230 ml, 550 ml, 630 ml. Give your answer in litres.
5. Subtract 3.54 kg from 4.23 kg. Give your answer in grams.
6. Multiply 65 cm by 6. Give your answer in metres.
7. A cat eats 1 tin of food a day. 1 tin contains 390 g of food. How much does the cat eat in a week? Give your answer in kilograms.
8. A bag of flour weighs 1.5 kg. A recipe for a chocolate cake requires 175 g of flour. How many cakes can be made from the one bag of flour?
9. A kitchen wall is 3.4 m long. A householder wants to fit the following kitchen units to the wall:

> Double cupboard unit – 1000 mm wide
> Single drawer unit   – 500 mm wide
> Single cupboard unit – 500 mm wide
> Fridge   – 600 mm wide

How much space will these units take altogether? Is there sufficient space left to fit a washing machine along the same wall if the washing machine is 600 mm wide?

Answers on page 169.

6

Ratios and proportions

Ratios

Suppose someone saved $\frac{1}{5}$ of their earnings and spent the remaining $\frac{4}{5}$, this would mean, for every £5.00 earned £1.00 would be saved and £4.00 would be spent. We say the ratio of saving to spending is £1.00 to £4.00; this is written £1.00:£4.00. Since the units of both parts of the ratio are the same this can be simplified to 1:4.

For every £10.00 earned the amount saved would be £2.00 (£2.00 is $\frac{1}{5}$ of £10.00), and the amount spent would be £8.00. This is a ratio of £2.00 to £8.00 or 2:8.

For every £100 earned the amount saved would be £20.00 and £80.00 would be spent. The ratio of saving to spending is £20.00 to £80.00 or 20:80.

All of these ratios represent the same proportion of saving to spending. So 1:4 = 2:8 = 20:80.

The ratio 2:8 can be simplified by dividing both 2 and 8 by 2 to give 1:4 and the ratio 20:80 can be divided by 20 to give 1:4.

The ratio 1:4 is written in its simplest form.

Example
Write the following ratios in their simplest form.
(a) 6:16   (b) 21 m:15 m

Solution
(a) 6:16   We can divide both numbers by 2 to give 3:8.

(b) 21 m:15 m Since the units are the same, this can be written as 21:15. Dividing both numbers by 3 we get 7:5.

Now try these:
Write the following ratios in their simplest form:
(a) 6:27   (b) 14 cm:7 cm   (c) £4:£20   (d) 35 Kg:10 Kg
(e) 42p:12p.

Answers on page 169.

Example
Reduce the ratio 20p:£1.20 to its simplest form.

Solution
In order to simplify this ratio we must first use common units.

$$20p:£1.20 = 20p:120p = 20:120$$

If we divide both numbers by 20 we get $20:120 = 1:6$.

Now try these:
(a) 45p:£1.25   (b) 30 cm:1 m   (c) 225 g:1.5 Kg

Answers on page 169.

Example
Three people contribute to a football pools stake each week. Alan gives £2.00, Bill gives £1.00 and Colin gives £3.00. How much should each get if they win a total of £5400?

Solution
It is fair for the winnings to be shared in the same proportion as the contributions. The three men contribute in the proportion 2:1:3. Altogether there are $2+1+3=6$ parts.

Alan contributes $\frac{2}{6}$
Bill contributes $\frac{1}{6}$
Colin contributes $\frac{3}{6}$

The winning total of £5400 should be shared out as follows:

Alan should get $\frac{2}{6} \times £5400 = £1800$

Bill should get $\frac{1}{6} \times £5400 = £900$

Colin should get $\frac{3}{6} \times £5400 = £2700$

(check: $£1800 + £900 + £2700 = £5400$)

Now try these calculations:
1. A charity dance raises £1800 which is to be donated to four charities A, B, C and D in the proportion of 5:1:3:6. How much money should each of the charities receive?
2. The following mixture of compounds is used in the preparation of a new lawn:

> Sulphate of Ammonia 2 parts
> Superphosphate   4 parts
> Bonemeal   4 parts
> Sulphate of Potash   1 part

If 2 kg of the mixture are required to cover a lawn calculate how much of each of the above compounds you would need to the nearest gram.

Answers on page 170.

Proportions

One of the most useful topics in arithmetic is proportions. If you can understand proportions you will be able to solve many different types of problem.

Direct proportions

Example
Find the cost of 5 packets of crisps at 18p per packet.

Solution

If 1 packet costs 18p
then 5 packets cost $5 \times 18p = 90p$

Now try these calculations
Find the cost of the following:
(a) 4 items at 35p each.   (b) 12 items at £2.00 each.

(c) 9 items at 45p each.   (d) 6 items at £1.23 each.

Answers on page 170.

Example
If a box of 12 greetings cards cost £2.40 find the cost of 1 card.

Solution

> If 12 cards cost £2.40, i.e. 240p
> then 1 card costs 240p/12 = 20p

Now try these:
Find the cost of 1 item in the following:
(a) 4 items cost 96p   (b) 7 items cost £2.24
(c) 5 items cost £12.50   (d) 12 items cost £3.60

Answers on page 170.

Now we can combine the two types of example and deal with more complicated problems.

Example
If you walk at 3 miles per hour how long would it take for you to walk 5 miles?

Solution
First work out how long it takes to walk 1 mile, then work out how long it takes to walk 5 miles.

> You walk 3 miles in   1 hour
> this is 3 miles in   60 minutes
> so you walk 1 mile in 60/3 minutes

and you walk 5 miles in $5 \times 60/3$ mins = 100 minutes = 1 hour 40 minutes.
Walking at 3 miles per hour, it would take you 1 hour 40 minutes to walk 5 miles.

Now try this calculation:
Mr Tebbs lives 8 miles from his place of work. He usually travels to work by bicycle and he cycles at an average speed of 12 mph.

(a) How many minutes does it take Mr Tebbs to get to work when he goes by bicycle?
(b) If Mr Tebbs had a lift to work in a car which made the journey at an average speed of 30 mph, how many minutes would he save compared with making the journey by bicycle?

Answers on page 170.

Inverse proportion

If 1 person is doing a task such as painting a wall, digging a garden, and so on, then several people sharing the work will take a shorter time to complete it. Similarly 1 person will take longer to complete a task than when it is done by several people. This type of problem is known as inverse proportion.

Example
If 1 person is able to clean all the windows in a house in 40 minutes how long would it take for 2 people to complete the task?

Solution
2 people will take half the time of 1 person to do the job.

1 person takes 40 minutes

so 2 people would take 40/2 minutes = 20 minutes

Example
If 3 people take 2 hours to paint the walls of a room how long would it take 1 person?

Solution
1 person will take 3 times as long to paint the walls.

3 people take 2 hours

so 1 person takes 3×2 hours = 6 hours

Now try these calculations:
1. If 1 person takes 5 hours to clean all of the carpets in a house how long would it take 2 people?

2. If 3 people take 7 hours to pave the drive of a new house how long would it take 2 people?

Answers on page 170.

Example
You engaged 3 men to dig the foundations to an extension to your house and they took 15 hours to complete the task. How long would it have taken 5 men?

Solution
First work out how long it would take 1 man, and then work out how long it would take 5 men.

$$3 \text{ men take} \quad 15 \text{ hours}$$
$$1 \text{ man takes } 3 \times 15 \text{ hours}$$

$$5 \text{ people take } \frac{3 \times 15}{5} \text{ hours} = 9 \text{ hours}$$

5 people would take 9 hours.

Example
A car travelling at 30 miles per hour takes 1 hour 12 minutes to complete a journey. How long would it take if the car travelled at 45 miles per hour?

Solution
At 30 mph the journey takes 1 hour 12 minutes = 72 minutes
at 1 mph the journey takes 30×72 minutes

$$\text{At 45 mph the journey takes } \frac{30 \times 72}{45} \text{ minutes} = 48 \text{ minutes}$$

At 45 mph the journey takes 48 minutes.

Now try these calculations:
1. A train travelling at 64 mph takes 2 hours 12 minutes to complete a journey. How long would the journey take if the train travelled at 88 mph?
2. An aircraft travelling at 600 mph takes 1 hour 40 minutes to make a journey. How long would the journey take if the aircraft travelled at 480 mph?

Answers on page 170.

7

Percentages

The word percent means per hundred, and all percentages are fractions of 100. We normally write percent as % which simply means /100.

Example

$$50\% \text{ means } 50 \text{ in every } 100 \text{ or } \tfrac{50}{100}, \text{ i.e. } \tfrac{1}{2}$$
$$5\% \text{ means } 5 \text{ in every } 100 \text{ or } \tfrac{5}{100}, \text{ i.e. } \tfrac{1}{20}$$
$$35\% \text{ means } 35 \text{ in every } 100 \text{ or } \tfrac{35}{100}, \text{ i.e. } \tfrac{7}{20}$$

NB: 100% means 100 in every 100 or $\tfrac{100}{100}$, i.e. 1

So if we say that 7% of the workforce are unemployed then we know that 7 persons in every 100 or $\tfrac{7}{100}$ are unemployed.

Changing percentages to fractions

We can see from the examples given above that in order to show a percentage as a fraction we divide it by 100:

Example

$$20\% = \tfrac{20}{100} = \tfrac{1}{5}$$
$$12\tfrac{1}{2}\% = \tfrac{12\frac{1}{2}}{100} = \tfrac{25}{200} = \tfrac{1}{8}$$

(Multiply the top and bottom of the fraction $\tfrac{12\frac{1}{2}}{100}$ by 2 then cancel down)

$$33\tfrac{1}{3}\% = \frac{33\tfrac{1}{3}}{100} = \frac{100}{300} = \frac{1}{3}$$

(Multiply the top and bottom of the fraction $33\frac{1}{3}/100$ by 3 then cancel down)

Now try these:
Convert the following percentages to fractions:
(a) 5%   (b) 20%   (c) 15%   (d) 60%   (e) 42%   (f) 35%
(g) 4%   (h) 84%   (i) $66\frac{2}{3}$%   (j) $17\frac{1}{2}$%   (k) $11\frac{1}{4}$%   (l) $26\frac{1}{3}$%

Answers on page 170.

Changing fractions to percentages

Since all percentage figures are fractions of 100 they give us a very useful measure to compare all kinds of statistics such as prices, wages, examination results, and so on. It is often useful to show simple fractions as percentages. In order to do this we need to change the fraction to a fraction of 100, this is done by multiplying the fraction by 100%. NB $100\% = \frac{100}{100} = 1$

Example
$$\frac{3}{5} = \frac{3}{5} + \frac{100}{1}\% = \frac{300}{5}\% = 60\%$$
$$1\frac{1}{4} = \frac{5}{4} \times \frac{100}{1}\% = \frac{500}{4}\% = 125\%$$
$$\frac{5}{8} = \frac{5}{8} \times \frac{100}{1} = \frac{500}{8}\% = 62\frac{1}{2}\%$$

Now try these:
Convert the following fractions to percentages:
(a) $\frac{1}{2}$ (b) $\frac{2}{5}$ (c) $\frac{1}{4}$ (d) $\frac{7}{10}$ (e) $\frac{5}{6}$ (f) $\frac{4}{15}$
(g) $\frac{24}{25}$ (h) $\frac{3}{50}$ (i) $1\frac{1}{2}$ (j) $\frac{123}{236}$ (k) $\frac{257}{183}$
(l) $\frac{563}{154}$

Answers on page 170.

Expressing one quantity as a percentage of another

Example
In a school 2 out of every 5 of the teachers are women. What percentage of the teachers are men?

Solution

If $\frac{2}{5}$ of the teachers are women then $\frac{3}{5}$ are men. To change $\frac{3}{5}$ to a percentage multiply by 100% as explained above.

$$\tfrac{3}{5}=\tfrac{3}{5}\times\tfrac{100}{1}\%=\tfrac{300}{5}\%=60\%$$

60% of the teachers are men.

Now try these calculations:
1. Express 4 as a percentage of 24.
2. A student in an examination gets 110 out of 150, express this mark as a percentage.
3. A small factory employs 45 people of whom 32 are unskilled, 6 are skilled, and 7 are clerical staff. What percentage of the employees are unskilled, skilled and clerical?
4. In a class of 30 children 5 of them were absent. What percentage of the class was present?

Answers on page 170.

Finding the value of a percentage of a quantity

Suppose we wish to calculate 24% of 35 m. This can be written:

$$\tfrac{24}{100}\times 35\,\text{m}=80.4\,\text{m}$$

Now try these:
(a) 4% of 75   (b) 34% of 52   (c) $17\frac{1}{2}$% of 120   (d) 87% of 215

Answers on page 170.

Example

A mail order firm gives its agents $12\frac{1}{2}$% commission on all the goods they sell. If an agent sells goods which are valued at £320 what commission can she claim?

Solution

$$12\tfrac{1}{2}\%\text{ of }320=\tfrac{12\frac{1}{2}}{100}\times 320=\tfrac{25}{200}\times\tfrac{320}{1}=40$$

The commission is £40.

Example

A firm of estate agents charge a fee for selling property at the rate of

$2\frac{1}{2}\%$ on the first £30 000; and then 2% on the remainder. What would the seller have to pay the estate agent if his house was sold for £75 000?

Solution
The fee consists of $2\frac{1}{2}\%$ of £30 000 and 2% of the remainder which is £45 000.

$$\text{The basic fee} = 2\frac{1}{2}\% \text{ of } 30\,000 + 2\% \text{ of } 45\,000$$
$$= (\tfrac{2\frac{1}{2}}{100} \times 30\,000) + (\tfrac{2}{100} \times 45\,000)$$
$$= (\tfrac{5}{200} \times 30\,000) + (\tfrac{2}{100} \times 45\,000)$$

The basic fee $= 750 + 900 = 1650$
The basic fee is £1650.

Example
A company's profits for last year were £97 428. If, this year, profits decreased by 14% what is the company's new profit figure?

Solution
$$14\% \text{ of } £97\,428 = \tfrac{14}{100} \times 97\,428 = £13\,639.92$$

The new profits are £97 428.00 − £13 639.92 = £83 788.08.

Finding the whole from a given percentage

Given that 35% of a quantity is £46, find 100%
If 35% is equivalent to £46 then 1% is equivalent to £46/35
so that 100% is equivalent to $100 \times £46/35 = £131.43$

Example
A dress is reduced by £8.10. If the reduction is 15%, what was the original price of the dress, and what is its sale price?

Solution
$$\text{Since 15\% is equivalent to £8.10}$$
$$\text{then 1\% is equivalent to £8.10/15}$$
$$\text{and 100\% is equivalent to } 100 \times £8.10/15 = £54.00$$

The original price of the dress was £54.00
The sale price $= £54.00 - £8.10 = £45.90$.

Example

A customer bought an item in a sale for £33.50. If the sale price of the item was 15% less than the original price what was the original price of the item?

Solution

Since there is a reduction of 15% then the sale price is $(100-15)\%$ which is 85% of the original price. Thus we know that £33.50 is 85% of the original price.

Since 85% is equivalent to £33.50
then   1% is equivalent to £33.50/85

100% (the original price) is equivalent to $100 \times £33.50/85 = £39.41$.

Now try these:

1. Calculate the following:
   (a) 30% of £4.00   (b) 56% of 25 g   (c) 4% of £12.50
   (d) 95% of 200   (e) $2\frac{1}{4}\%$ of 17.6 lb   (f) 15% of 16 m

2. Decrease the following:
   (a) £5 by 15%   (b) £430 by $12\frac{1}{2}\%$ (c) 75 g by $33\frac{1}{3}\%$
   (d) 4 km by 25%   (e) 141 by 80%   (f) £25 000 by 45%

3. Find the whole quantity of which:
   (a) £3.00 is $2\frac{1}{2}\%$ (b) £630 is 95%   (c) 40p is 6%
   (d) 5 g is 17%   (e) 8 cm is 62%   (f) 14 ml is 36%

4. In a firm a medical officer found that 57% of the employees do not smoke. What percentage of the employees smoke?

5. In an examination students are awarded grades A, B, C, D, E. E represents a fail. What percentage of the students failed if the following grades were awarded: 4% of students had grade A, 17% grade B, 34% grade C, 29% grade D?

6. A brand of drinking chocolate consists of sugar, cocoa and an additive. 25% of the drinking chocolate is cocoa. (a) What percentage of the chocolate is made up of sugar and the additive?   (b) What is the combined weight of sugar and the additive in a 250 g tin of drinking chocolate?

7. A group of workers get a pay increase of 9%. If each worker's gross income was £160.35 per week, what will an individual be paid after the award?

8. The price of a product is 86p. What will its new price be if: (a) The price is increased by 6%? (b) The price is decreased by 4%?

9. A woman saves 14% of her net monthly income; and spends the remainder. If she spends £640 in a month what is her total net monthly income?

10. A leading soap manufacturer offers 10% extra soap powder free with the large size packet. If the 'special officer' packet weighs 924 g, what is the weight of a normal packet?

11. Two boxes each contain twenty-five oranges. In one box two of the oranges are bad, and in the second box three are bad. What percentage of the fifty oranges are bad?

12. In a batch of eggs two were rotten. If this is 8% of the total, how many eggs were there altogether in the batch?

13. A car purchased three years ago cost £7500 and has depreciated in value by 10% each year since then. What is its current value?

14. These are the marked prices of some items in a jewellers: Wrist watch ... £59.50; Earrings ... £17.50; Necklace ... £37.95. The shop advertises that it is reducing all marked prices by 15%. Calculate the reduced price of each item listed.

15. An Electricity Board announces a $12\frac{1}{2}$% increase in all its prices. Calculate the new cost of the following quarterly bills: (a) £56.48   (b) £98.40   (c) £135.24.

16. A man's earnings are £125.00 a week and he saves 16% of this sum each week. Calculate his total savings after one year.

17. A basket of groceries costs £56.65. (a) What will it cost in two years time if the rate of inflation is 9% per annum?   (b) What would the same groceries have cost two years ago if the rate of inflation had been 10% each year?

Answers on page 170–171.

8

Measuring lengths, areas and volumes

Measuring perimeters

The measurement round the edge of a given figure or shape is called
the perimeter. We meet such measurement quite frequently in
gardening or in DIY projects.

Example
What is the perimeter of the following reactangle?

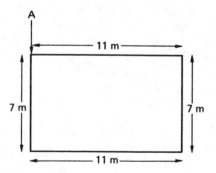

Solution
Starting at A and going round the rectangle in a clockwise direction
we have 11 m + 7 m + 11 m + 7 m = 36 m.

The perimeter of the rectangle is 36 m.

Example
A builder wishes to place stakes for fencing around the edge of a rectangular building plot at distances of 2 m apart. The sides of the plot measure 110 m, 104 m, 110 m, 104 m. How many stakes will be required?

Solution
The perimeter of the plot = 110 m + 104 m + 110 m + 104 m = 428 m. The number of stakes required at two metres apart: = 428/2 = 214 stakes.

Now try the following:
1. Find the perimeters of the following figures:

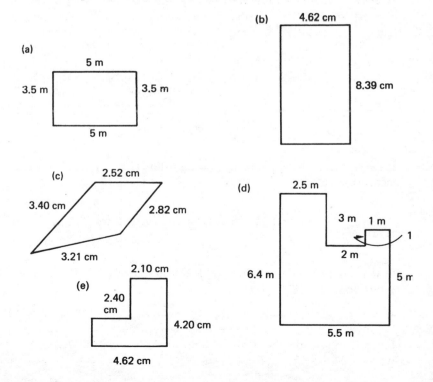

2. Assume you want to plant roses around the edge of a lawn.
   (a) If the lawn measures 9 m by 12 m, and the planting distance between the roses is 0.5 m, how many rose bushes should you buy?   (b) What will the total cost of the rose bushes be if each bush costs £2.25?   (c) If you decide that this is too expensive and you buy £135 worth of rose bushes, how far apart must each rose bush be planted?

Answers on page 171.

Measurement of an angle

An angle measures the amount or degree of 'turning' between two straight lines. The length of the lines does not affect the size of the angle.

Angles are measured in degrees; the sign is °. For example, the angle in Fig. 1 is 45°, and the angle in Fig. 2 is 135°.

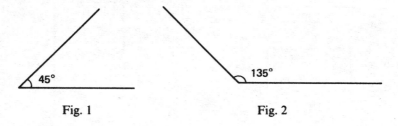

Fig. 1   Fig. 2

If two arms of the angle lie in a straight line (Fig. 3) then the angle measures 180°.

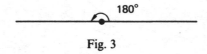

Fig. 3

If the angle goes full circle back to the first arm (Fig. 4), the angle measures 360° (180° + 180°).

360°

Fig. 4

An angle which is met in everyday activities is the corner of a square or rectangle. This angle occurs when one arm has gone round one quarter of a circle (Fig. 5) and measures 90°.

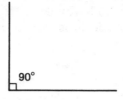

90°

Fig. 5

An angle of 90° is called a right-angle.

How to measure an angle

Angles are measured with a protractor. Place the protractor on one of the lines of the angle so that the centre of the bottom line of the protractor coincides with the point of the angle, and then read the angle on the protractor's scale (see Fig. 6).

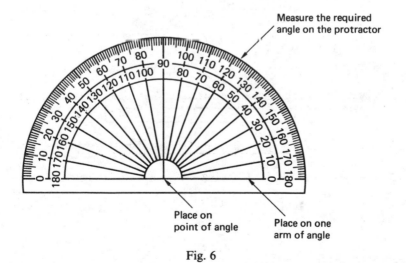

Measure the required angle on the protractor

Place on point of angle

Place on one arm of angle

Fig. 6

You can see from Fig. 6 that the protractor measures to only 180°. To measure an angle which is greater than 180° then you have to apply the rule that all angles round a point add up to 360°. The smaller angle remaining can then be measured with the protractor and this is then subtracted from 360°.

Constructing angles

The protractor can be used to construct angles. The technique is illustrated in Fig. 7, where an angle of 50° has been constructed.

Example
Construct an angle of 50°.

Solution

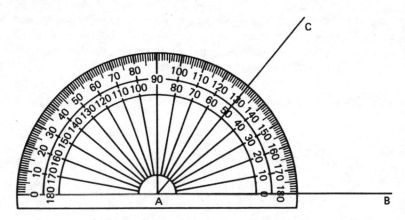

Fig. 7 Diagram of construction

To construct an angle of 50°, firstly draw a straight line A – B, then place the centre of the bottom line of the protractor at one end of the line at A and mark the angle which is required at C. A and C are then joined.

Now try the following:
Using a protractor construct the following angles:
(a) 30°   (b) 110°   (c) 57°   (d) 76°   (e) 270°

Area

Area measures the space which is enclosed by lines. For example, Fig. 8 is a rectangle which has a length of 4 cm and a width of 3 cm.

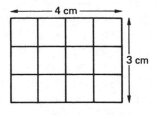

Fig. 8

The diagram shows that the rectangle can be divided into 12 squares with sides of 1 cm. The area of the rectangle is, therefore, 12 square centimetres. Square centimetres can be abbreviated to sq cm or cm^2. This area can be found simply by multiplying each side of the rectangle:

$$3 \text{ cm} \times 4 \text{ cm} = 12 \text{ sq cm}$$

NB: 12 sq cm is NOT the same as a square with each side equal to 12 cm.

Similarly if we have a square with each side equal to 5 m; from Fig. 9 we can see that the total area of the square can be divided into 25 squares with each side equal to 1 metre.

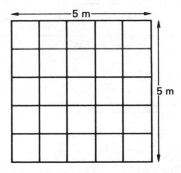

Fig. 9

The area of the total square can be found by multiplying the two sides:

so 5 m × 5 m = 25 square metres.

This is abbreviated to 25 sq m or 25 m².

We see, then, that the area of a rectangle or square can be found by multiplying the length by the width. The length and width should be measured in the same units.

Now try these:
The following table gives the length and width of three rectangles A to C. Calculate the area of each rectangle:

	length	width
A	7.00 m	3.00 m
B	12.42 cm	9.00 cm
C	20.00 cm	0.5 m (change this to cm)

Answers on page 171.

In everyday affairs it is often necessary to calculate areas which are more complicated than squares or rectangles. However, in many instances the calculation can be simplified by converting the areas into squares or rectangles.

Example
Fig. 10 shows the layout of a new kitchen, and we need to know the L shaped shaded area in order to calculate how many floor tiles to order.

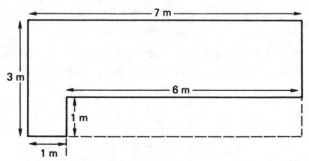

Fig. 10

Solution

One method is to calculate the total area of the kitchen and then deduct the area of the rectangle which is not shaded. The remainder then gives the area of the L shaped space.

$$\text{Area of the kitchen} = 7\,m \times 3\,m = 21\,sq\,m$$
$$\text{Area not required} = 6\,m \times 1\,m = 6\,sq\,m$$

Area to be covered = 21 sq m − 6 sq m = 15 sq m.

Conversions into squares and rectangles can be made even when the area to be calculated is complicated.

Example

Calculate the total area of the following shape:

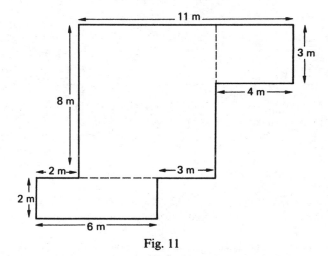

Fig. 11

Solution

The area can be divided into three rectangles shown by the dotted lines. The total area is the sum of the areas of the three rectangles:

$$A = 2\,m \times 6\,m = 12\,sq\,m$$
$$B = 7\,m \times 8m = 56\,sq\,m$$
$$C = 3\,m \times 2\,m = 6\,sq\,m$$

$$\text{Total area} \quad 74\,sq\,m$$

Now try the following:

1. Calculate the shaded areas of the following three shapes:

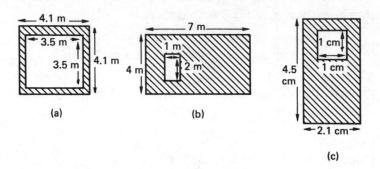

(a)   (b)   (c)

2. Find the area of the following figures:

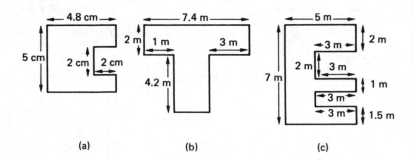

(a)   (b)   (c)

Answers on page 171.

Area of the walls of a room

In order to calculate the area of a 'box' shape, such as a room, it is useful to imagine the shape opened out.

Example
The room in the following diagram has these dimensions:

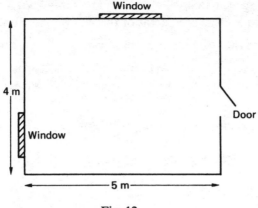

Fig. 12

The height of the room is 3.5 m. The door is 0.8 m × 1.9 m. Each window is 1.1 m × 1.8 m. Calculate the total area of the walls of the room.

Solution
The following diagram shows the 'box' opened out.

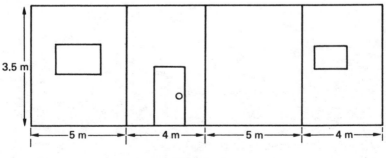

Fig. 13

We can now calculate the total wall area of the room by finding the sum of the lengths of each wall, and then multiplying this sum by the height of the room. The area of the doors and windows is then deducted to give the total area of the walls.

Add the length of each wall $= 5\,m + 4\,m + 5\,m + 4\,m = 18\,m$
Multiply this by the height $= 18\,m \times 3.5\,m = 63\,sq\,m$

Deduct the area of door and windows:

$$door = 0.8\,m \times 1.9\,m = 1.52\,sq\,m$$
$$window\ 1 = 1.1\,m \times 1.8\,m = 1.98\,sq\,m$$
$$window\ 2 = 1.1\,m \times 1.8\,m = 1.98\,sq\,m$$

Total area of doors and windows $= 5.48\,sq\,m$

Total net area of the walls of the room is

$$63\,sq\,m - 5.48\,sq\,m = 57.52\,sq\,m.$$

The circle

The edge of a circle, that is, the line which actually draws the circle, is called the circumference.

The radius of a circle is the line from the centre of a circle to its circumference; and it does not matter where on the circumference the radius is drawn, it is always the same length for any particlar circle.

The diameter of a circle is a line drawn through the centre which divides the circle into two halves.

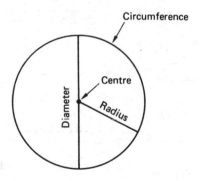

Fig. 14

From fig. 14 we can see that the diameter of a circle is, and always will be, double the length of the radius.

The radius, in association with a value called pi (π), is used to calculate both the area of a circle and the circumference of a circle.

Pi (π), pronounced 'pie', is always the same value which is approximately 3.142 or $\frac{22}{7}$.

The circumference of a circle

It is difficult to measure the circumference of a circle directly, but a value for it can be found by multiplying the diameter by π (3.142). This gives us the formula:

$$\text{Circumference} = \pi \times \text{diameter}$$

This is sometimes written as $C = \pi \times d$

Since the diameter is $2 \times$ radius, then this gives us another formula for the circumference

$$\text{Circumference} = \pi \times 2 \times \text{radius}$$

This is sometimes written as $C = 2 \times \pi \times r$

Example
Calculate the circumference of a circular carpet which has a diameter of 3 metres.

Solution

$$\text{Circumference} = \pi \times \text{diameter}$$
$$\text{Circumference} = 3.142 \times 3.0\,\text{m} = 9.426\,\text{m (3 d.p.)}$$
$$= 9.4\,\text{(1 d.p.)}$$

Now try the following:
Here are the diameters of five circles. Find the circumference of each circle and give your answers to 2 d.p.:
(a) 3 cm   (b) 7 m   (c) 5.2 cm   (d) 15.03 cm   (e) 26.74 cm

Answers on page 171.

To find the area of a circle

The area of a circle is $\pi \times$ radius $\times$ radius. This is sometimes written as:

Area of circle $= \pi \times r \times r$ where r is the radius, *or*
Area of circle $= \pi \times r$ squared where r squared means $r \times r$.

Example
Find the area of a flower bed which has a radius of 2.4 metres.

Solution

$$\begin{aligned}
\text{Area of a circle} &= \pi \times r \times r \\
&= 3.142 \times 2.40 \,\text{m} \times 2.40 \,\text{m} \\
&= 18.1 \,\text{sq m (1 d.p.)}
\end{aligned}$$

Now try the following:
The radius of each of five circles is given, (a) to (e). Find the area of each circle.

(a) 4 cm   (b) 6.9 m   (c) 10.5 cm   (d) 110 cm   (e) 12.8 m

Answers on page 171.

9

Reading tables and charts

Information can be presented in a variety of ways, for example tables, dials, charts and diagrams.

Tables

Tables are used to present information such as travel timetables, tariffs, and repayment of loans. The basic elements of a table are rows and columns. Care has to be taken to read accurately the information corresponding to the required row and column.

Example

Vehicle Tariffs: Ferry Crossing.

Mr Adams wishes to travel on the Ferry Crossing with a car and caravan. The length of the caravan, including the tow bar is 4.5 m. Using the table overleaf calculate how much he would have to pay for the vehicles at night-time on the 20th August.

Solution
Mr Adams is taking two vehicles: a car and a caravan. These are found in row 1 and row 3.

He is travelling at night in the peak season. From the table we see that this is column 4.

Vehicle Tariffs: single journey (£'s)			Standard Season 1 Jan → 15 Jun 16 Sep → 31 Dec		Peak Season 16 Jun → 15 Sep	
Category			Day	Night	Day	Night
1	Cars and Other Vehicles Maximum length 5.5 metres Maximum height 2.4 metres, inc. roof rack etc.	Each	30.00	40.00	35.00	50.00
2	Cars, Minibuses and Mobile Homes If not applicable to category 1 above	Price per metre or part thereof	8.00	14.00	10.00	17.00
3	Caravans Length calculation must include towbar	Price per metre or part thereof	9.50	14.00	17.00	17.00
4	Car Trailers Maximum length 5.5 metres Maximum height when loaded 2.4 metres Length calculation must include towbar	Price per metre or part thereof	8.00	14.00	17.00	17.00
5	Light Commercial Vehicles If not applicable to categories 1, 2, 3 or 4 Maximum length 6.5 metres	Price per metre or part thereof	16.00	16.00	18.00	18.00
6	Coaches	Price per metre or part thereof	5.00	10.00	5.00	10.00
7	Motorcycles/Scooters	Each	7.00	13.00	9.00	16.00
8	Motorcycles with Sidecars	Each	10.00	20.00	12.00	24.00
9	Bicycles/Mopeds	Each	2.00	5.00	3.00	7.00

The cost of the car is row 1, column 4 which is £50.00.

The caravan is row 3 column 4 which gives £17.00 per metre. Rounding up we find that the length of the caravan is 5 metres. So the cost of the caravan is $5 \times £17.00 = £85.00$

The total cost $= £50.00 + £85.00 = £135.00$.

Now try these calculations:

Using the table of vehicle tariffs calculate the cost of transporting the following vehicles on the ferry:

1. Motorcycle on 1st October in the day.
2. Car with trailer which is 3.4 metres long on 4th March at night.

Answers on page 171.

Reading timetables

Many travel timetables use the 24 hour clock. Since it is important to be accurate about travel information, it is essential to be familar with the 24 hour clock.

The diagram shows the times on a 24 hour clock. The numbers inside the face show the times from 12 o'clock midnight to 12 noon. The numbers on the outside show times from 12 noon to midnight, i.e. the afternoon and evening.

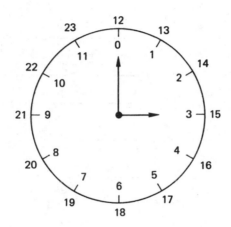

So for example,

> 1500 hours is 3 o'clock in the afternoon
> 2100 hours is 9 o'clock at night
> 0800 hours is 8 o'clock in the morning

NB: The 24 hour clock is sometimes shown (as in this example) as 15.00 hours, 21.00 hours, 8.00 hours. The 'full stop' is not a decimal point but simply separates the hours from the minutes.

> 4 pm is 16.00 hours
> 10 pm is 22.00 hours
> 2.32 pm is 02.32 hours (2 hrs 32 mins)

To convert the 24 hour clock to the 12 hour clock the rules are:

> 00.00 is midnight

01.00 to 12.59 are the same (morning)
13.00 to 23.59 – subtract 12 (afternoon to midnight)

Example

Convert the train departure times shown in bold on the following timetable, which are expressed in 24 hour clock terms, to 12 hour clock terms:

		1	1	1	1	1	1	1	1	1
HIRING CROSS	dep	**12 00**	16 00	17 15	18 30	20 45	21 05	**21 56**	21 56	23 05
PARKWAY	dep	12 48L	—	—	—	—	—	—	—	—
Whiteway	dep	12 55L	—	—	—	—	20 55 L	—	—	23 56L
Biddle Bridge	dep	**13 09**	16 07P	—	19 14P	20 53	21 12	21 14P	22 15	**00 23**
Fairfield	arr	—	16 30	—	—	—	21 33	—	—	—
Badgersholt	arr	—	16 32	—	—	—	—	—	—	—
Miles Green	arr	—	16 38	—	—	—	—	—	—	—
BEECHEND	arr	13 34	17 42	18 45	19 02	21 21	21 41	21 58	23 01	01 10

HIRING CROSS	dep	**02 43**
Whiteway	dep	—
Biddle bridge	dep	03 00
Fairfield	arr	—
BEECHEND	arr	04 24

Solution

$$21.56 - 12.00 = 9.56 \text{ pm}$$
$$13.09 - 12.00 = 1.09 \text{ pm}$$
$$02.43 \qquad\quad = 2.43 \text{ am}$$
$$12.00 \qquad\quad = 12.00 \text{ pm}$$
$$00.23 + 12.00 = 12.23 \text{ am}$$

Now try the following:

Convert the following 24 hour times to 12 hour times:
(a) 15.38   (b) 22.05   (c) 12.12   (d) 00.00   (e) 03.21

Answers on page 171.

The system for converting 12 hour times to 24 hour clock times is:

12 o'clock midnight is 00.00
1.00 am to 12.59 are the same (morning)
1.00 pm to 11.59 pm – add 12 (afternoon to midnight)

Example

Convert the following 12 hour times to 24 hour clock times:
10.43 am   1.56 pm   12.03 am   6.05 pm   12.32 pm

Solution

$$10.43 \text{ am} \qquad = 10.43$$
$$01.56 \text{ pm} + 12.00 = 13.56$$
$$12.03 \text{ am} - 12.00 = 00.03$$
$$06.05 \text{ pm} + 12.00 = 18.05$$
$$12.32 \text{ pm} \qquad = 12.32$$

Now try these

Convert the following 12 hour times to 24 hour clock times:
(a) 9.32 am (b) 12.02 pm (c) 3.45 pm (d) 11.59 pm
(e) 12.30 am

Answers on page 172.

Dials and Guages

One of the most common dials are electricity meters which show the consumption of electricity in a house.

Example

Shown below are the positions of the dials on an electricity meter at the end of a quarter. Write down the meter reading.

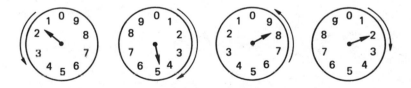

Solution

The first dial shows that over 1 but not 2 has been used
The second dial shows that over 4 but not 5 has been used
The third dial shows that over 8 but not 9 has been used
The fourth dial shows exactly 2
Note the direction in which the indicator of each dial travels!

The reading is therefore 1482.

Weight scales

Since weights in everyday use range from ounces to stones or from grams to kilograms there is a wide variety of scales in use. The important point to remember is to note the units used by the scales and to note the division of the units.

Example
The following weight scales gives both imperial and metric weights: metric is the outer scale and imperial the inner. Write down the weight shown.

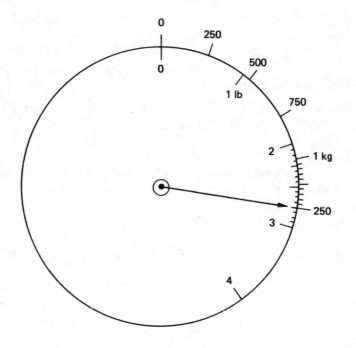

Solution
Imperial – There are 16 divisions for each lb shown and therefore each small division is equal to one ounce. The reading shown is: 2 lb 12 oz.

Metric – There are 10 small divisions between 1 kg and the next 250 grams so each small division is equal to 25 grams. The reading shown is 1 kg 250 g or 1.250 kg.

Diagrams

Diagrams are frequently used to display information. If they are constructed correctly they allow the reader to read the information easily, quickly and accurately. Unfortunately diagrams are sometimes used to mislead and deceive the reader. It is useful to be aware of the correct way of interpreting diagrams.

When you read a diagram it is important first to read any labels or notes so that you understand what the diagram is trying to show, and secondly make sure you know what units of measurements are being used.

There are many different forms of diagram, the most common ones are dealt with below.

Graphs

Graphs show the relationship between two sets of data. They have two lines: a horizontal line and a vertical line to represent the data. These lines are called axes. The axes should be clearly labelled and the units of measurement should be shown.

Example
The graph given on the next page shows the relationship between height and weight of women. The horizontal axis gives the weight in stones, and the vertical axis gives the height in feet and inches.

A woman who wants to check whether she is overweight should take a straight line across from her height and a line up from her weight and put a mark where the two lines meet. This will indicate which region of the graph she is in.

Example
Using the graph find the weight category of a woman who weighs 11 stones and is 5′ 5″ tall.

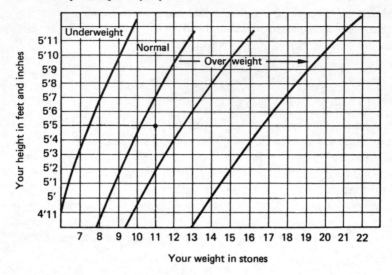

Your weight in stones

Solution

Take a straight line across from 5′ 5″ and a line up from 11 stone these two lines meet in the overweight region of the graph. The woman is, therefore, overweight.

Now try these:

1. Using the graph find the weight category of a woman who weighs 7.5 stones and is 5′ 6″ tall.
2. What is the ideal weight range of a woman who is 5′ 7″ tall?

Answers on page 172.

Misleading graphs

In the graph given above the two axes start at 4′ 10″ and 6 stones respectively rather than from zero. This is sensible in this example. Unfortunately omitting part of the axis can often give a misleading picture.

Example

Consider the following information:

Company Sales (Number of items)
1989: 60 000
1990: 65 000

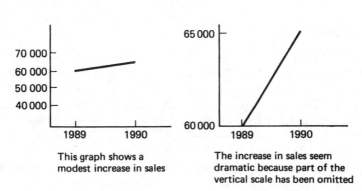

| This graph shows a modest increase in sales | The increase in sales seem dramatic because part of the vertical scale has been omitted |

Pie diagrams

Pie diagrams, as the name implies, look like a pie cut into different sectors. The size of each sector is proportional to each value it represents.

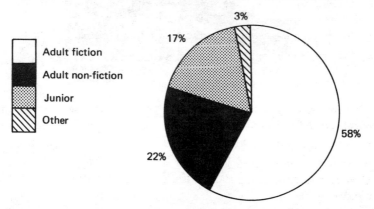

Public libraries: percentage of books borrowed

The pie diagram on the previous page shows that most books issued in libraries are adult fiction.

The number of junior books issued is only 17%.

The smallest category is the 'other' category, which includes computer software, sound and video recordings.

Bar charts

There are many different types of bar chart from simple bar charts to rather more complicated component bar charts. The main feature of all bar charts is that the length of the bars indicate the size of data. With any bar chart it is important to read any notes and the key to make sure that you know how the data is being represented.

Example

The following is a chart which shows two simple bar charts which are superimposed on one another. One of the charts gives the average daily temperature in London and the other gives the average daily temperature at a holiday resort.

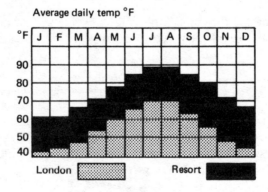

From the chart we can see that in August, for example, the temperature in London is 70 degrees Fahrenheit and at the resort it is almost 90 degrees Fahrenheit.

SECTION TWO

Everyday applications of arithmetic

Section Two is practical and uses lots of everyday examples and so many of the calculations do not work out neatly or exactly. In many instances, such as estimating, it is not important to arrive at the exact answer and approximations can be used. However, it is important that you understand when and why you are using approximate answers; Section Two will help you to improve your understanding.

10

Cooking

There is a wide range of arithmetic skills which are used when we cook. Ingredients have to be weighed and measured, we have to calculate proportions and to read scales and guages; and we often have to convert the numbers given in a recipe to suit our own requirements.

Cooking conversion

Most modern recipes are expressed in both imperial and metric measurements, but sometimes only one standard is given and it can be necessary to convert from imperial measures to metric or vice versa.

The tables in the Appendix give some of the more common conversions (see page 180). The conversion figures given in the appendix are the most accurate ones; however, when cooking they can be tedious and it is often easier to use the following approximations for small quantities:

$$1 \text{ oz} = 25 \text{ g}$$
$$1 \text{ pint} = 600 \text{ ml}$$
$$1 \text{ teaspoon} = 5 \text{ ml spoon}$$
$$1 \text{ tablespoon} = 15 \text{ ml spoon}$$

Example
Convert the following recipe ingredients from metric to imperial.

Ox tongue and orange salad

500 g cooked ox tongue
1×2.5 ml spoon salt
4×15 ml spoons olive oil
1×15 ml spoon grated lemon rind.

Solution
To convert 500 g of ox tongue we could use the table in the appendix and we see that 50 g is equivalent to 1.76 oz

so 1 g is equivalent to 1.76/50 oz
500 g is equivalent to $500 \times (1.76/50)$ oz $= 17.6$ oz

17.6 oz is very approximately 1 lb.
Since the other measures are small they can be converted using the simpler conversions given above. We now get:

1 lb cooked ox tongue
$\frac{1}{2}$ teaspoon salt
4 tablespoons olive oil
1 tablespoon lemon rind.

Now try this calculation:
Convert the following recipe ingredients from metric to imperial.

Potato salad

500 g potatoes
120 ml mayonnaise

1 × 15 ml spoon lemon juice
1 × 2.5 ml spoon salt
2 × 15 ml spoons chives
4 × 15 ml spoons chopped leeks.

Convert the following recipes from imperial to metric.

Chicken pimento

½ oz butter
7 oz can pimentos
4 oz cheese
5 fl oz double cream
¼ pint milk.

Answers on page 173.

Temperature conversion

A second area of cooking where conversion skills are important is in the measurement of temperature. Cooking temperatures can be measured on three scales: Gas marks, Fahrenheit and Celsius (Centigrade).

The table for converting Celsius to Fahrenheit is in the Appendix, page 181.

Food quantities

The number of servings given by recipes varies, but the most common are 4 and 6. Quite often, however, we may need to cater for servings other than 4 and 6. In such cases the amounts given in the recipes have to be changed using proportions.

Example

The following recipe for Irish stew pie serves 4 people. Find the quantities required to serve 3 and 17 people.

Irish stew pie

425 g stewing lamb	50 g onions
400 g potatoes	200 g carrots
100 g celery	

Solution

First divide the quantities by 4 to obtain the amount required for one serving, then multiply this by 3 to obtain recipe quantities for three servings:

From 4 to 3 servings:

$$\tfrac{3}{4} \times 425\,\text{g} = 318.75\,\text{g stewing lamb}$$
$$\tfrac{3}{4} \times 400\,\text{g} = 300\,\text{g potatoes}$$
$$\tfrac{3}{4} \times 100\,\text{g} = 75\,\text{g celery}$$
$$\tfrac{3}{4} \times \;\;50\,\text{g} = 37.5\,\text{g onions}$$
$$\tfrac{3}{4} \times 200\,\text{g} = 150\,\text{g carrots}$$

Solution from 4 to 17 servings:

$$\tfrac{17}{4} \times 425\,\text{g} = 1806.25\,\text{g stewing lamb}$$
$$\tfrac{17}{4} \times 400\,\text{g} = 1700\,\text{g potatoes}$$
$$\tfrac{17}{4} \times 100\,\text{g} = 425\,\text{g celery}$$
$$\tfrac{17}{4} \times 50\,\text{g} = 212.5\,\text{g onions}$$
$$\tfrac{17}{4} \times 200\,\text{g} = 850\,\text{g carrots}$$

Now try this:

The following recipe is designed to serve 6 people. What quantities are needed to serve 2 or 15 people?

Cheese and mushroom pie

9 oz shortcrust pastry	0.75 pints milk
2.25 oz butter	6 oz grated cheese
1.20 oz cornflour	4 oz mushrooms
1.5 teaspoons curry powder	3 tomatoes

Answers on page 173.

Cooking in bulk

The introduction of domestic freezers has meant that many people cook in bulk in order to store the cooked food. The method of converting recipes from small to large amounts is the same as the method used in the previous section.

Example

The following list of ingredients are those required to make 1 chocolate cake. Calculate the quantities required to make 3 cakes.

Chocolate cake

175 g butter	4 eggs
175 g caster sugar	30 ml milk
5 ml vanilla essence	175 g self-raising flour
100 g plain chocolate	25 g cocoa powder

Solution

To make 3 cakes we need 3 times the quantities:

Butter	$3 \times 175\,g = 525\,g$	Eggs	$3 \times 4 = 12$
Caster sugar	$3 \times 175\,g = 525\,g$	Milk	$3 \times 30\,ml = 90\,ml$
Vanilla essence	$3 \times 5\,ml = 15\,ml$	Flour	$3 \times 175\,g = 525\,g$
Plain chocolate	$3 \times 100\,g = 300\,g$	Cocoa powder	$3 \times 25\,g = 75\,g$

Now try these:

1. The following ingredients for shortbread biscuits are sufficient for 12 biscuits: 150 g flour; 50 g caster sugar; 100 g margarine. Calculate the ingredients required for 30 biscuits and 50 biscuits.
2. You have invited 15 people to dinner. Calculate how much meat you will need to buy to serve 12 oz to each person.
3. The following recipe for Chicken casserole is designed for 4 people. Calculate the quantity of ingredients you would need to serve 18 or 25 people.

Chicken casserole

2 onions	1 oz butter
3 sticks celery	4 chicken joints
$\frac{1}{4}$ lb mushrooms	3 tablespoons flour
2 oz bacon	$\frac{3}{4}$ pint chicken stock
1 tablespoon of oil	15 oz cans tomatoes

Answers on page 174.

Cooking times

The time required to cook meat varies according to the type. The following list gives the approximate time per lb which is usually recommended when cooking various joints of meat.

Beef	20 mins per lb + 20 minutes
Lamb	30 mins per lb + 30 minutes
Gammon	25 mins per lb + 20 minutes
Chicken	20 mins per lb + 20 minutes
Pork	25 mins per lb + 20 minutes

A common calculation which is often necessary is the estimate of how much time to allow to cook a particular weight of meat.

Example
You have to serve a meal at 7.15 pm and decide to cook a chicken joint. The meal includes a 3 lb 10 oz chicken, and it will take 30 minutes to prepare. What time should you start?

Solution
It is simpler to do this calculation if the weight is expressed completely in pounds:

$$10 \text{ oz is equivalent to } \tfrac{10}{16} \text{ lbs}$$
$$\tfrac{10}{16} \text{ lbs} = 10 \div 16 = 0.63 \text{ lbs}$$

Therefore 3 lb 10 oz is equivalent to 3.63 lbs.

Time required to cook the chicken joint is:

$$3.63 \times 20 \text{ mins} + 20 \text{ mins 'cooking time'} = 72.6 \text{ mins} + 20 \text{ mins}$$
$$= 92.6 \text{ mins}$$

Preparation time = 30 mins. Therefore total time = 92.6 + 30 = 122.6 minutes or 2 hours 2.6 minutes.

The time to start preparing the meal is just over 2 hours before you want to eat and therefore you should begin the preparation just before 5.15 pm.

Now try these:
1. For Sunday lunch you are to prepare beef which weighs 3.2 lbs. If the preparation time is 25 minutes at what time should you start preparing the meal if it is to be served at 12.30 pm?

2. What time should a piece of pork be put in the oven to cook, if it weighs 5 lb 5 oz, for it to be ready for 8.30 pm? The recipe suggests that you should allow 25 minutes per pound plus 20 minutes.

3. Allowing 30 minutes per pound + 30 minutes find the total cooking times of the following weights of meat:
   (i) 3 lbs   (ii) 4 lbs   (iii) 2.5 lbs   (iv) 3 lbs 8 oz   (v) 2 lbs 10 oz   (vi) 3 lbs 2 oz   (vii) 4.75 lbs

Answers on page 174.

Weight loss during cooking

Frequently food loses a weight during preparation and cooking. For instance vegetables are often stripped of their outer leaves and fruit may be cored and peeled; similarly the weight of a joint of meat may be reduced by boning and trimming. In all these instances there can be further loss of weight due to cooking.

An arithmetic problem which can occur in cooking is to estimate the serving weight of food.

Example

It is estimated that a joint of meat, which weighs 2.5 kg before cooking, will lose 18% of its weight after cooking. What weight will be available for serving?

Solution

The original weight is 2.5 kg.
Weight loss = 18% of 2.5 kg = $\frac{18}{100} \times 2.5$ kg = 0.45 kg

The estimated final weight is $2.5 - 0.45 = 2.05$ kg.

Now try this:

Mrs Crabb picks 7.8 lbs of apples from her orchard and peels, cores and cooks them. If the finished weight of the apples after cooking was 6.3 lbs what percentage of the fruit was lost during preparation and cooking?

Answer on page 174.

Quite often the problem given above is reversed: we decide what weight of food we want to be served and, given an estimate of the weight loss during preparation and cooking, we have to calculate the weight of food to be purchased.

Example
At a dinner for 8 you want to serve 200 g portions of meat to each of the guests. If the expected weight loss due to preparation and cooking is 32% what weight of meat should be purchased?

Solution
The total requirement for cooked meat is $8 \times 200\,g = 1600\,g$

Since there is a weight loss of 32% then
1600 g represents 68% (100 − 32) of what is required

1% of the weight of the meat to be purchased is

$$1600/68\,g = 23.53\,g$$

Therefore 100% of the weight of the meat is $100 \times 23.53\,g = 2353\,g$ or 2.35 kg

Therefore 2.35 kg of meat should be purchased.

Now try this:
You have been asked to help to prepare for a small wedding party to which 20 guests have been invited. It is intended to purchase the equivalent of 300 g of meat for each guest.
  (i) What is the total weight of the meat to be purchased?
 (ii) It turns out that about 18% of the meat is lost during the preparation. Calculate the weight of meat which goes into the oven.
(iii) It is estimated that a further 12% of the meat will be lost during cooking. What will be the total weight of the meat after cooking?
(iv) Calculate the weight of each portion if all of the meat is shared equally between the 20 guests.

Answers on page 174.

11

Shopping

Unit costs

Comparisons between the prices of similar products can be difficult if they are sold in different weights, volumes, sizes, and so on. The way to overcome the problem is to standardise the price by calculating a unit cost for each product. A unit cost is the price of a product expressed in a single unit.

Example
The following table gives the prices of three different sizes of the same brand of cereals. Calculate which size is the best value for money.

$$500\,g \text{ cost } £0.96$$
$$750\,g \text{ cost } £1.25$$
$$2\,kg \text{ cost } £2.99$$

Solution
Convert each of the three sizes to unit costs of 1 g

500 g cost £0.96 (96p)   **so** 1 g costs 96/500p = 0.192p
750 g cost £1.25 (125p) **so** 1 g costs 125/750p = 0.167p
2 kg cost £2.99 (299p)   **so** 1 g costs 299/2000p = 0.15p

2 kg size is the best value.
You can use standard cost calculations to compare prices of different shops.

Example

The following table gives the price of wine (in £s) in 3 different sizes from 5 different shops:

	A	B	C	D	E
75 cl	2.60	2.65	2.40	2.50	2.60
1 litre	3.44	3.80	3.40	3.22	3.40
1.5 litres	5.10	5.20	4.95	4.75	5.00

Convert the prices to a standard cost to determine which shop sells the cheapest wine (per unit).

Solution

The most common size wine bottle is 0.75 litre so make this the standard unit.

To calculate the standard unit for shop A:

1 litre (100 cl) costs £3.44 so 75 cl cost $75/100 \times £3.44 = £2.58$
1.5 litres (150 cl) cost £5.10 so 75 cl cost $75/150 \times £5.10 = £2.55$

This method of calculation gives the following prices per 0.75 litre for all five shops. A 1.5 litre bottle, for instance, contains 2×75 cl.

	A	B	C	D	E
75 cl	2.60	2.65	2.40	2.50	2.60
1 litre	2.58	2.65	2.55	2.42	2.55
1.5 litres	2.55	2.60	2.48	2.38	2.50

The cheapest bottle per unit is 1.5 litres from shop D which cost £2.38 in 0.75 litre units.

Now try this:

The following chart gives the price (in £s) of 5 different shampoos in different size bottles. Calculate the unit cost of each to determine which is the cheapest per unit.

	A	B	C	D	E
50 ml	0.40	0.55	0.60	0.40	0.50
110 ml	0.90	0.95	1.00	0.85	1.05
125 ml	1.00	1.10	1.10	1.05	1.15

Answer on page 174.

Package size conversion

A complication with the packaging of products is that they are sometimes sold in different weight or volume units: lbs and kilograms, pints and gallons or litres. The example given below shows one method of dealing with this problem.

Example
Fresh orange juice is supplied by the milkman at 42p per pint bottle. A similar product can be purchased in 1 litre cartons at the local supermarket for 72p. Which is the cheaper?

Solution
1 litre = 1.76 pints   1 litre cartons cost 72p
                          so 1.76 pints would cost 72p
                          so 1 pint would cost $72/1.76 = 41p$

The orange juice in the supermarket is 1p per pint cheaper.

VAT

Value Added Tax is charged against most goods and services. Since the prices of many goods are advertised or displayed for sale with VAT excluded it is useful to be able to increase a price by a given percentage. (Currently VAT is 15%).

Example
Increase the following prices by 11% and 15%:

$$£13.20 \qquad £4.95 \qquad £53.20$$

Solution
$£13.20 + 11\%$ of $£13.20 = £13.20 + \frac{11}{100} \times £13.20 = £13.20 + £1.45$
$= £14.65$

$£13.20 + 15\%$ of $£13.20 = £13.20 + \frac{15}{100} \times £13.20 = £13.20 + £1.98$
$= £15.18$

$£4.95 + 11\%$ of $£4.95 = £4.95 + \frac{11}{100} \times £4.95 = £4.95 + £0.54 = £5.49$

$£4.95 + 15\%$ of $£4.95 = £4.95 + \frac{15}{100} \times £4.95 = £4.95 + £00.74 = £5.69$

£53.20 + 11% of £53.20 = £53.20 + $\frac{11}{100}$ × £53.20 = £53.20 + £5.85
= £59.05

£53.20 + 15% of £53.20 = £53.20 + $\frac{15}{100}$ × £53.20 = £53.20 + £7.98
= £61.18

Now try these:
Increase the following prices by 9% and 17%:

(a) £3.91   (b) £6.23   (c) £31.43   (d) £207.61

Answers on page 174.

Discounts and reductions

It is useful to be able to calculate percentage reductions for a wide variety of percentages. During seasonal 'Sales', prices are often reduced by a percentage amount. Similarly reductions offered by club discounts or mail order agencies are often expressed in terms of a percentage.

Example
A product is normally sold at £26.50. Calculate its sale price if the following percentage reductions are made:

$$15\% \qquad 33.3\% \qquad 60\%$$

Solution
£26.50 − ($\frac{15}{100}$ × £26.50) = £26.50 − £3.98 = £22.52

£26.50 − ($\frac{33.3}{100}$ × £26.50) = £26.50 − £8.83 = £17.67

£26.50 − ($\frac{60}{100}$ × £26.50) = £26.50 − £15.90 = £10.60

Now try these:
A firm advertises audio cassettes at the following discount rates:

Quantity Ordered	Percentage reduction %
0–10	0.0
11–25	5.0
26–50	12.5
51–100	15.0
101+	22.0

The full price of each cassette is £0.40
Calculate the total cost of the order for each of the following quantities allowing for the discounts offered.
(a) 7   (b) 19   (c) 48   (d) 81   (e) 247

Answers on page 175.

12

Motoring

The cost of owning a car is high and, for many people, it is a relatively difficult figure to calculate. It is made up of running costs such as petrol, oil and tyres, ownership costs such as insurance, maintenance costs and depreciation.

Running costs

The most obvious expense in running a car is the cost of petrol. This particular cost can be difficult to predict. A calculation which helps us to make a fairly accurate forecast of running costs is the average petrol consumption of a car.

The average petrol consumption is the number of miles a car will travel, on average, for each gallon of petrol.

Example
A motorist keeps a record of the amount of petrol purchased and at the end of a 6 month period he finds that he has used 174 gallons, and has travelled 5690 miles. What is the average petrol consumption of the car?

Solution
Average petrol consumption = 5690/174 = 32.7 miles per gallon (mpg).

The average petrol consumption figure can be used to calculate the cost of individual journeys, or to predict the cost of petrol consumption over a period of time.

Example

A motorist intends to have a touring holiday in Scotland during which he anticipates that he will cover between 1750 and 1900 miles. If the average petrol consumption of his car is 28.6 mpg, and the predicted cost of petrol is £1.86 calculate the minimum and maximum cost of the petrol required for the holiday.

Solution

28.6 miles requires 1 gallon
so 1750 miles requires 1750/28.6 = 61.2 gallons

Minimum cost of petrol for 1750 miles = 61.2 × £1.86 = £113.80
The petrol required for 1900 miles = 1900/28.6 = 66.4 gallons
Maximum cost of petrol for 1900 miles = 66.4 × £1.86 = £123.50.

Now try this:

A shop supervisor plans to use her car to get to work, a return journey of 16 miles. The average petrol consumption of her car is 31.7 mpg, and she anticipates that the cost of petrol will be £1.83 per gallon. Assuming that the supervisor works a 5 day week, that she is not absent from work and takes a total of 6 weeks annual holiday make an estimate of the cost of the petrol used in a year.

Answer on page 175.

Servicing and maintenance costs

The cost of petrol is not the only cost in running a car: the durability of tyres and exhausts and the consumption of oil are largely determined by how often the car is used and how it is driven as well as by time.

A motorist can get a reasonable estimate of the cost of wear and tear on a car by recording regularly all bills for servicing and repair. Such bills have three parts: the cost of spare parts, the labour cost and VAT; it can be assumed that as a car becomes older then the costs will increase.

Example

The following table shows the cost of servicing and repair paid by a motorist over a three year period.

	£	£	£
	Year 1	**Year 2**	**Year 3**
Spares	103.25	289.36	221.05
Labour	115.23	175.46	111.17
Sub total	218.48	464.82	332.22
+ VAT (15% of sub total)	32.77	69.72	49.83
Total	251.25	534.54	382.05

1. If we assume that spares costs are likely to increase by 12% in the fourth year and labour costs by 8%, calculate the estimated costs for the fourth year.
2. If the motorist had carried out his own repairs and maintenance during the first three years how much would have been saved if no labour costs had been paid and spares could have been obtained at a discount of 20% (VAT can be ignored).

Solution (1)

	Year 3	**Year 4**	£
Spares	221.05	221.05 + (12% of 221.05)	
		221.05 + (12/100 × 221.05) = 247.58	
Labour	111.17	111.17 + (8% of 111.17)	
		111.17 + (8/100 × 111.17) = 120.06	
Sub totals	332.22		367.64
+ VAT	49.83		55.15
Total	382.05		422.79

Solution (2)
If the motorist were to carry out servicing and repairs himself he would incur the cost of the spares less 20%.

The total cost of spares in the three years:
£103.25 + £289.36 + £221.05 = £613.66

Deduct discount of 20%:
£613.66 − (20% of £613.66)
= £613.66 − (20/100 × £613.66) = £490.93

Saving on discount spares $= £613.66 - £490.93$
$$= £122.73$$
Saving on labour $= £115.23 + £175.46 + £111.17$
$$= £401.86$$
Total saving in three years is $£122.73 + £401.866$
$$= £524.59$$

Now try this:
Using the information in the previous example calculate the cost of running a car in the fourth year assuming that the cost of spares increase by 7% and labour costs by 5%.

Answer on page 175.

Cost per mile comparisons

There are two main types of expense in running a car: fixed costs which are not dependent on usage such as hire purchase or loan costs, insurance and road tax, and variable costs such as petrol, servicing and repairs. Once these expenses have been determined then we can make a calculation of average cost per mile for different mileages.

Example
The following table shows the estimated costs of running a family car for a particular year. Use the information to calculate the cost per mile for 6000, 9000, 12 000 and 15 000 miles.

Main costs:

	£
Hire purchase	850.00
Insurance	95.00
Road tax	80.00
AA	40.00
Total	1065.00

Running costs:

Servicing (6000 miles) $= £148.20$
Petrol per gallon $= £1.80$

It is estimated that the car does 28.6 miles to the gallon.

Solution

If we drove for a total of 6000 miles in a year we would incur:

(i) The main costs of £1065.00

(ii) The cost of petrol which would be:

number of gallons used $= 6000/28.6 = 209.8$ gallons
cost $= 209.8 \times £1.80 = £377.64$

(iii) The service costs $= £148.20$

The total cost is then divided by the total mileage to get the average cost per mile:

$$£(1065.00 + 377.64 + 148.20)/6000 = £0.27$$

In order to calculate the cost per mile for different mileages, using the method shown above, we can construct a table which shows the different total costs and then divide this sum by the respective mileage.

Miles	6000	9000	12 000	15 000
	£	£	£	£
Main costs	1065.00	1065.00	1065.00	1065.00
Petrol	377.64	566.43	755.24	944.06
Servicing	148.20	222.30	296.40	370.50
Total	1590.84	1853.73	2116.64	2379.56
Average cost	0.27	0.21	0.18	0.16

Now try this:

Using the information in the previous example calculate the total cost and the average cost per mile for 24 000 miles.

Answer on page 175.

Depreciation: cost of ownership

Apart from a house, a motor car is the most expensive item most families will purchase. Like many items of equipment cars depreciate in value. The extent of the depreciation is influenced by several factors such as the age of the car, how well it has been maintained, its mileage and its popularity with second hand car buyers. Consequently it can be difficult to estimate the future value of a car.

The most simple method of calculating depreciation is to estimate how many years the car is expected to last before it has no value, and this then gives an average annual cost.

Example

A new car costs £8000 and is expected to last for 10 years. Calculate the annual depreciation and show the estimated value of the car at the end of each year.

Solution

The car is expected to last 10 years and therefore it depreciates by £800, i.e. 8000/10 each year.

Year	Depn £	Value £
1	800	7200
2	800	6400
3	800	5600
4	800	4800
5	800	4000
6	800	3200
7	800	2400
8	800	1600
9	800	800
10	800	000

This method of calculating depreciation is easy to calculate, but the problem is that it is not realistic to assume that a car will depreciate in value evenly each year: most cars lose most of their value in the first few years of life.

The following method reflects this process more accurately: an estimate is made of the percentage of annual depreciation, and this percentage is then deducted from the car's value at the end of each year.

Example

A new car costing £8000 is expected to depreciate at 12% per annum. Show the year end estimated value of the car for 10 years.

Solution

The first year's depreciation will be 12% of £8000

$$= 12/100 \times £8000 = £960$$

The value at the end of the first year will be:

$$£8000 - £960 = £7040$$

The second year's depreciation will be 12% of £7040

$$= 12/100 \times £7040 = £845$$

The value at the end of the second year will be:

$$£7040 - £845 = £6195$$

The following table of depreciation for 10 years illustrates how the value of the car falls more rapidly when it is new. Compare the table with the table in the previous method.

Year	Depn £	Value £
1	960	7040
2	845	6195
3	743	5452
4	654	4798
5	576	4222
6	506	3716
7	446	3270
8	392	2878
9	345	2533
10	304	2229

The two methods of depreciation which are described above can be used to calculate the cost of owning any type of car.

Example

A second-hand car is estimated to have a remaining life of 4 years and is valued at £2850. Construct a table to show the annual cost of owning the car using the second method of depreciation.

Solution

If the car has an estimated life of 4 years then the annual cost of depreciation will be approximately $\frac{1}{4}$ or 25% of £2850. The cost of the first year will be:

$$(\tfrac{25}{100}) \times £2850 = £713$$

Year	Depn 25% £	Value £
1	713	2137
2	534	1603
3	401	1202
4	301	901

You will notice with this method of calculating depreciation that the value of the car can never be £0; and this is generally a realistic view. You should remember that calculating the depreciation on a car gives us a reasonable estimate of the cost of car ownership, it is unlikely that the calculation will give the accurate second-hand value of your car.

Now try this:

A second-hand car is estimated to have a remaining life of 6 years and is valued at £5620. Construct a table to show the annual cost of owning the car using the second method of depreciation.

Answers on page 175.

Motor insurance

The cost (premium) of motor insurance depends on several factors including the value of the vehicle, the type of insurance cover and the record and experience of the driver.

A driver who does not make a claim over a period of time is often given a 'no claims bonus' or discount up to a maximum of 65% on the premium. In most circumstances this discount is withdrawn if a claim is made.

A calculation which can be made in the event of an accident is the cost of repair against the cost of losing a 'no claims bonus'.

Example

A motorist's insurance premium is £180 per annum and this is after a 'no claims bonus' of 60% has been allowed. The bonus is 15% per annum built up over 4 years of trouble-free motoring. The motorist is involved in a minor accident and the estimated cost of repair to his damaged car is £200.

Should the motorist claim on his insurance if then his 'no claims bonus' is reduced to 30%? If he does not make a claim then the bonus of 60% remains intact. If he does claim then he has to pay £25.00 of the cost of repair, as well as losing two years bonuses.

Solution
Since the motorist has a discount of 60% of the annual premium then the premium of £180 is 40% of the total.

$$1\% \text{ of the total premium is } £180.00/40 = £4.50$$
$$100\% \text{ of the total premium is } £4.50 \times 100 = £450.00$$

The cost of losing the first year's bonus is:

$$30\% \text{ of } £450.00 = £135.00$$

The cost of losing the second year's bonus is:

$$15\% \text{ of } £450.00 = £67.50$$

In addition to losing these two year's bonuses the motorist has to pay £25.00 towards the cost of repair.

The total cost of making the claim is:

$$£135.00 + £67.50 + £25.00 = £227.50$$

The cost of making the insurance claim exceeds the estimated cost of repair by £227.50 − £200 = £27.50, therefore it's probably better not to make a claim.

Now try these:
The estimated cost of repair to a car involved in an accident is £190.00. Assuming that the insurance company requires the first £25.00 of any claim to be met by the motorist, calculate the difference between the estimated cost of repair with the cost of losing the 'no claims bonus' in the following two cases.

1. The actual premium is £240.00 after a 'no claim bonus' of 60% of the total premium has been deducted. A minor claim on the

insurance policy would reduce the bonus to 30% in the first year and 45% in the second year.

2. The actual premium is £578.00 after a 'no claims bonus' of 15% has been deducted. A claim would mean that no bonus would be allowed.

Answers on page 175.

13

Travelling

Reading charts

A useful skill which is required when we are planning routes and journeys is that of reading charts and diagrams. Most travel charts are designed so that they are easy to read.

Example

The following mileage chart shows the distances in miles between principal towns in Great Britain. Find the distances from:

> Cardiff to Edinburgh
> Bristol to Dover
> York to Dover

Solution

Determine which box on the chart corresponds to both towns. For instance, to find the distance from Cardiff to Edinburgh look down the column of Cardiff until it corresponds with the row for Edinburgh; we then have Cardiff to Edinburgh; we then have Cardiff to Edinburgh = 368 miles.

Similarly: Bristol to Dover = 187 miles and York to Dover = 264 miles.

Distance chart (mileage between British towns).

	Aberdeen	Birmingham	Bristol	Cambridge	Cardiff	Carlisle	Coventry	Dover	Edinburgh	Exeter	Glasgow	Gloucester	Harwich	Hull	Leeds	Liverpool	Manchester	Newcastle-on-Tyne	Northampton	Norwich	Nottingham	Oxford	Plymouth	Preston	Sheffield	Southampton	Stoke-on-Trent	York
Birmingham	405																											
Bristol	484	87																										
Cambridge	446	100	143																									
Cardiff	488	102	43	174																								
Carlisle	212	193	272	256	275																							
Coventry	417	18	91	82	114	209																						
Dover	559	182	187	112	225	369	164																					
Edinburgh	119	286	365	327	368	93	303	440																				
Exeter	559	162	218	347	167	242	440																					
Glasgow	142	287	365	350	370	94	303	463	44	441																		
Gloucester	452	52	35	117	56	239	57	176	332	110	334																	
Harwich	509	159	188	67	228	323	141	112	393	242	417	172																
Hull	341	124	203	123	225	150	111	236	222	279	245	169	185															
Leeds	310	109	194	208	146	110	257	191	269	210	189	211	55	73														
Liverpool	330	90	160	174	164	117	107	269	210	235	212	128	128	73	40	35												
Manchester	329	79	159	154	172	117	94	255	210	235	211	241	221	40	35	128												
Newcastle-on-Tyne	225	200	284	227	298	57	202	340	106	359	143	249	295	91	153	134	119	210										
Northampton	429	50	128	50	128	236	31	137	309	177	330	72	124	134	119	153	173	215	112									
Norwich	475	156	208	62	236	284	138	153	356	177	330	72	110	117	124	184	256	57	124									
Nottingham	372	50	137	84	152	182	48	197	253	213	378	179	143	66	73	67	97	70	41	57	94							
Oxford	466	64	69	80	105	257	50	129	346	276	179	66	102	156	97	154	142	247	139	139	254	180						
Plymouth	601	204	118	260	160	389	209	284	482	350	213	151	151	321	160	311	311	401	100	219	254	180	302					
Preston	300	105	184	188	87	123	283	42	138	276	125	183	61	38	73	163	94	148	214	100	169	302	281	68				
Sheffield	342	77	163	116	178	145	78	229	245	91	149	108	56	37	85	64	81	135	179	50	37	65	146	233	192			
Southampton	530	128	75	129	115	320	114	139	411	105	415	91	199	246	89	190	173	217	148	214	148	65	241	331	65	170		
Stoke-on-Trent	364	43	124	123	138	152	57	219	199	245	246	89	152	108	56	37	38	163	73	179	50	80	52	47	99	77		
York	303	127	214	151	229	113	126	264	184	289	207	179	217	38	24	97	81	135	111	122	78	172	57	211	159	236	147	
London	492	110	116	54	153	298	92	72	373	170	392	105	73	168	197	184	273	65	65	111	122	57	211	159	77	99	147	196

Now try these:
Using the mileage chart find the mileage from:

    (a) Bristol to Glasgow   (c) Edinburgh to Cambridge
    (b) Dover to Norwich   (d) Sheffield to Cambridge

Answers on page 175.

Calculating journey distances

The mileage chart can be used to find the total mileage of a particular journey.

Example
Mr and Mrs Briggs plan to travel from Exeter to York. They decide that initially they will travel to see their son in Cambridge; and then go to Leeds to see some friends before eventually going to York. Using the mileage chart find the total mileage for their intended journey, and calculate how much shorter the journey would have been if they had travelled directly from Exeter to York.

Solution

$$
\begin{array}{ll}
\text{Exeter to Cambridge} & = 218 \text{ miles} \\
\text{Cambridge to Leeds} & = 144 \text{ miles} \\
\text{Leeds} & = 24 \text{ miles} \\
\hline
\text{Total} & = 386 \text{ miles}
\end{array}
$$

From the chart: Exeter to York = 289 miles

The difference between the two journeys is:

$$386 \text{ miles} - 289 \text{ miles} = 97 \text{ miles}.$$

Now try the following:
Mr Grey lives in London and is planning to travel by car directly to Liverpool; however, he intends to visit Bristol on the return journey. Calculate the total mileage for the round trip.

Answer on page 175.

Petrol consumption

The amount of petrol used on a journey not only depends on the make of the car and the size of the engine, but it will also vary with such factors as variations in speed and the number of stops and starts. By keeping records of journeys made we can calculate the average petrol consumption of a vehicle.

Example
Imagine that you made a journey of 315 miles and that you estimate that the car used 8 gallons of petrol. Calculate the average petrol consumption of the car.

Solution
To get the average petrol consumption we divide the number of miles travelled by the number of gallons used:

$$= 315/8 = 39.38 \text{ miles per gallon}$$

This means on average we were able to travel 39.38 miles on one gallon of petrol.

We do not always know how many gallons of petrol a car has used and we have to calculate this by dividing how much we have spent on petrol by the price per gallon of petrol.

Example
During the month of November Miss Pugh recorded the following motoring information:

Petrol Purchases £	
5 Nov	14.50
13 Nov	17.90
19 Nov	15.00
26 Nov	18.56
Total	65.96

Mileage record on 1st Nov   = 34 763
Mileage record on 30th Nov = 35 742

If the average price of a gallon of petrol was £1.84 calculate the average petrol consumption.

Solution

£1.84 buys 1 gallon

so £1.00 buys 1/1.84 gallons

Miss Pugh spent £65.96 and so the number of gallons she bought was $65.96 \times 1/1.84 = 35.85$ gallons.

Total miles travelled $= 35\,742 - 34\,763 = 979$ miles.

Average petrol consumption is the total miles travelled divided by the number of gallons bought:

$= 979/35.85 = 27.3$ miles per gallon.

Now try the following:
The table gives information on the petrol purchases for a car during May:

Petrol Purchases £	
2 May	13.90
10 May	16.75
16 May	18.00
22 May	16.00
28 May	16.89

The mileage record of the car at the beginning of May was 23 987, and at the end of May was 25 393; and the average price of petrol during May was £1.86. Calculate the car's average petrol consumption.

Answer on page 175.

Estimating journey costs

If we know the average petrol consumption of a car then we can calculate, approximately, how much petrol is required for a particular journey, and, given the cost of petrol, how much the journey will cost.

Example
If, on average, a car does 35.4 miles to the gallon estimate how much petrol will be used to travel from Carlisle to Glasgow.

If the price of petrol is £1.85 per gallon calculate the estimated cost of travelling to Glasgow and back from Carlisle.

Solution
Using the mileage chart on page 103 the distance from Carlisle to Glasgow is 94 miles. The number of gallons required for the single journey is the total number of miles divided by the average petrol consumption:

$$= 94/35.4 = 2.66 \text{ gallons.}$$

The total gallons required for the round trip:

$$= 2 \times 2.66 \text{ gallons} = 5.32 \text{ gallons.}$$

The estimated cost of the journey to Glasgow and back from Carlisle is $5.32 \times £1.85 = £9.84$.

Now try the following:
If, on average, a car does 27.8 miles to the gallon estimate how much petrol will be used to travel from Cardiff to Newcastle. If the price of petrol is £1.83 per gallon calculate the estimated cost of the return journey.

Answer on page 175.

Average speed

The time taken to make a journey is not simply dependent on the distance. If our starting point and destination are located near to the same motorway then the journey generally will be quicker than if we have to travel across country. However we can make reasonable estimates of the average speed of specific journeys by keeping records.

Example
A family, which lives in Liverpool, plans to spend a holiday in a small village near Bristol. At the beginning of the holiday they travel directly to Bristol via the motorway, but at the end of the holiday they return home through the Welsh border country.

The following record gives details of their routes and the start time and finish time. Using the information calculate the average speed of each part of the total trip.

Route	Distance (miles)	Start Time	Finish Time
Liverpool→Bristol M6/M5	161	8.15	11.22
Bristol→Chepstow M4	16	9.12	9.34
Chepstow→Hereford A466	33	9.34	10.42
Hereford→Shrewsbury A49	53	11.15	12.51
Shrewsbury→Liverpool A41	71	12.51	2.49

Solution
The average speed is found by calculating how long each part of the trip took in hours and dividing the number of miles travelled by this sum.

Liverpool to Bristol took from 8.15 to 11.22 which is:

$$3 \text{ hrs } 7 \text{ mins or } 3 \text{ and } 7/60 \text{ hours} = 3.12 \text{ hours}$$

The average speed $= 161/3.12 = 51.6$ mph.

Similarly Bristol to Chepstow took from 9.12 to 9.34 which is:

$$22 \text{ mins or } 22/60 \text{ hours} = 0.37 \text{ hours}$$

The average speed $= 16/0.37 = 43.2$ mph.

Now try these:
1. Calculate the average speed for each of the remaining parts of the journey to Liverpool.
2. The following information shows the route and times taken of a journey from Sheffield to Blackpool. Calculate the average speed for each stage of the journey.

Route	Distance (miles)	Start Time	Finish Time
Sheffield→Manchester A57	37	12.46	1.37
Manchester→Preston M6	31	1.59	2.25
Preston→Blackpool A583	17	2.30	3.08

Answers on page 176.

Journey times and average speed

If we know the average speed of a journey, that is the average number of miles travelled in one hour, then, given the distance to be travelled we can calculate how long a journey will take.

Example

If a car travels at an average speed of 45 miles per hour estimate how long it would take to travel from Harwich to Stoke-on-Trent.

Solution

From the chart Harwich to Stoke-on-Trent = 190 miles. Since, on average, the car travels 45 miles in one hour then the time taken to cover 190 miles is:

190/45 hours = 4.22 hours or 4 hours 13 minutes.

Now try the follow calculations:

1. How long would it take you to travel from Southampton to Oxford at an average speed of 38 miles per hour?
2. How much longer would the journey take if you called at London on the way?

Answers on page 176.

Scale

Because maps have to represent precise distances in the real world, all of them are drawn using a particular relationship between distances on the map and on the ground. This relationship is called 'scale'. A map of North America would probably use a scale of 1 cm on the map to represent 50 miles on the ground. On the other hand a map of England or Scotland would probably use 1 cm on the map to represent 10 miles on the ground. Many road maps use a scale in the region of 1 cm is equivalent to 2 miles.

All maps give the scale which they are using and so we can easily estimate the mileage shown on any map by measuring the distance on the map and converting this distance into miles.

Example
If the scale of a map is 3 cm is equivalent to 5 miles convert the following measurements of the map to miles.
(a) 6.2 cm   (b) 28.9 cm   (c) 145.2 cm

Solution
(a) 3 cm is equivalent to 5 miles so 1 cm is equivalent to 5/3 miles.
   6.2 cm is equivalent to $5/3 \times 6.2$ miles = 10.33 miles.
(b) 28.9 cm is equivalent to $5/3 \times 28.9$ miles = 48.16 miles.
(c) 145.2 cm is equivalent to $5/3 \times 145.2$ miles = 242.0 miles.

Now try the following:
If the scale of a map is 4 cm = 10 miles convert the following measurements of the map to miles:
(a) 3.6 cm   (b) 12.3 cm   (c) 178.5 cm

Answers on page 176.

Travelling abroad

The main arithmetic activity when travelling abroad is conversion from one unit to another: mainly from metric to imperial. The most common metric units which will be used by the British traveller abroad are kilometres and litres.
The conversion factors are:

   1 kilometre is equivalent to $\frac{5}{8}$ miles
   1 mile is equivalent to $\frac{8}{5}$ kilometres.

Example
A tourist, on disembarking at Calais made the following journey:

   Calais to Paris   –   293 km
   Paris to Geneva   –   571 km
   Geneva to Vienna – 1029 km

Convert each stage of the journey into miles to the nearest mile.

Solution
Calais to Paris   – 293 km is $\frac{5}{8} \times 293$ miles = 183 miles
Paris to Geneva – 571 km is $\frac{5}{8} \times 571$ miles = 357 miles

Geneva to Vienna – 1029 km is $\frac{5}{8} \times 1029$ miles $= 643$ miles.

Now try this calculation:
When on holiday in Italy you hire a car and record the following route:

$$\text{Milan to Venice} = 276 \text{ km}$$
$$\text{Venice to Rome} = 535 \text{ km}$$
$$\text{Rome to Milan} = 569 \text{ km}$$

Convert each stage of the route from kilometres to miles, to the nearest mile.

Answer on page 176.

Example
The average petrol consumption of a car is 34 mpg. Express this in terms of kilometres per litre (kpl).

Solution
$$34 \text{ miles} = 34 \times 1.6 \text{ km} = 54.4 \text{ km}$$
$$1 \text{ gallon} = 4.55 \text{ litres}$$

Petrol consumption per litre $= 54.4/4.55 = 11.96$ kpl.

Now try this calculation:
The average petrol consumption of a car is 10.5 kilometres per litre. Express this in miles per gallon.

Answer on page 176.

Exchange rates

The most common form of conversion when travelling abroad is converting £ to other forms of currency. Exchange rates are constantly changing, and they are not necessarily the same within a country. The following table shows typical exchange rates for 5 common currencies:

£1.00 will buy the following currencies:

$$10.75 \quad \text{Francs} \quad \text{(France)}$$
$$1.805 \quad \text{Dollars} \quad \text{(United States)}$$

2326.00 Lire   (Italy)
   3.145 Marks   (Germany)
 202.00 Pesetas (Spain)

Example
Using the table of exchange rates convert the following prices into
sterling:
(a) 1254 Francs   (b) 223 Dollars   (c) 62 Marks

Solution
(a) 1 Franc = £1/10.75
so 1254 Francs = 1254 × £1/10.75 = £116.65

(b) 1 Dollar = £1/1.805
so 223 Dollars = 223 × £1/1.805 = £123.55

(c) 1 Mark = £1/3.145
so 62 Marks = 62 × £1/3.145 = £19.71

Now try these calculations:
Using the table of exchange rates convert the following prices into
sterling.
(a) 42 879 Lire   (b) 14 350 Pesetas   (c) 895 Dollars

Answers on page 176.

14

Working

Conversion of basic income

Calculation of pay is common to all working people, and the various deductions for such things as taxation and national insurance can make the calculations complicated. Earnings before deductions are made are known as gross income and sometimes we have to convert weekly or monthly payments to annual gross income.

Example
Given a salary of £6420 a year calculate the gross monthly salary.

Solution
Divide £6420 by 12:

$$\text{Monthly salary} = 6420/12 = £535 \text{ per month.}$$

Example
If you earn £4890 a year what is your gross average weekly wage?

Solution
Divide 4940 by 52:

$$\text{Weekly wage} = 4940/52 = £95.00 \text{ per week.}$$

Example
If, on average and before any deductions are made, you earned £128.59 a week what is your gross annual income?

Solution

Multiply £128.59 by 52:

Gross annual income = 52 × £128.59 = £6686.68

Now try these:

1. If you earn £189.20 per week gross how much would you earn in a year?
2. Given a gross annual income of £8561 calculate the gross monthly income.
3. How much would you have to earn each month to obtain a gross annual income of £7500?

Answers on page 176.

Conversion of hourly payments

Many jobs do not pay a weekly wage: the pay is based on an hourly rate. In these cases the gross income is calculated by multiplying the number of hours worked by the rate for the job.

Example

A person working in a restaurant is paid £3.50 an hour. If, on average, he works 47 hours a week what is his gross weekly wage?

Solution

Multiplying £3.50 by 47:

Gross weekly wage = 47 × £3.50 = £164.50 per week

Example

A single parent has a part-time job which pays £47.03 a week. If the person has to work 16.5 hours in a week what is the hourly rate of pay?

Solution

Divide £47.03 by 16.5:

16.5 hours of work pays £47.03
so 1 hour of work pays £47.03/16.5

Hourly rate = 47.03/16.5 = £2.85 an hour.

Now try these calculations:

1. If you have a job which pays £4.85 an hour and you work for 39 hours in a week what is your gross weekly wage?
2. A part-time clerk is paid £85.50 for 30 hours work in one week. Calculate the hourly rate of pay.
3. How many hours in a week would you have to work to earn £103.50 if the hourly rate of pay was £4.50?

Answers on page 176.

National Insurance payments

The Government make deductions from people's wages for National Insurance and Income Tax. The amount which is deducted varies according to Government policy, but the following examples illustrate the basic principles.

Example

The following table gives the National Insurance payments which employees had to make in 1989/90 if they had not contracted out of the State Pension Scheme.

Weekly Earnings £	Contributions £
Below 41.00	Nil
41.00 – 69.99	5%
70.00 – 104.99	7%
105.00 – 154.99	9%
155.00 – 305.00	9%
Over 305.00	9% of £305.00

Calculate the National Insurance payments for the following weekly earnings:

(a) £48.50 (b) £194.00 (c) £500

Solution

(a) The contribution on weekly earnings of £48.50 is 5% of £48.50.

$$5\% \text{ of } £48.50 = \frac{5}{100} \times £48.50 = £2.43$$

(b) The contribution on weekly earnings of £194.00 is 9% of £194.00.

$$9\% \text{ of } £194.00 = \tfrac{9}{100} \times £194.00 = £17.46$$

(c) The contributions on weekly earnings of £500 is 9% of £305.00.

$$9\% \text{ of } £305 = \tfrac{9}{100} \times 305.00 = £27.45$$

Now try these:
Using the table of National Insurance payments calculate how much would be deducted from the following weekly wages:
(a) £58.90   (b) £98.20   (c) £133.00   (d) £456.70

Answers on page 176.

Taxation

Payments for taxation are slightly different from National Insurance contributions in that some allowances are deducted from our earnings before we have to pay tax. We then pay on what is known as 'taxable income'.

Example
The annual allowance for Miss Walsh, who is a single person, is £2605; if she earns £9400 in a year what is her taxable income?

Solution
Miss Walsh's taxable income = £9400 − £2605 = £6795.

Now try this:
If the married man's allowance is £4095 what is the taxable income of a married man earning £13 580?

Answer on page 176.

Income tax is deducted from our taxable income at different rates and the following table gives the range of tax which had to be paid on taxable income in 1988/89:

Taxable Income £	Rate %	Tax
First 19 300	25	4825
Remainder	40	—

Example
Calculate how much income tax would have to be paid on the following taxable incomes:
(a) £10 900   (b) £21 260

Solution
(a) The amount of tax due on £10 900 is 25%

$$25\% \text{ of } £10\,900 = 25/100 \times £10\,900 = £2725$$

(b) The amount of tax due on £21 260 is:

25% of £19 300 which is given in the table as £4825 and 40% of the remainder which is £21 260 − £19 300 = £1960.

$$40\% \text{ of } £1960 = 40/100 \times £1960 = £784$$

The income tax due on £21 260 = £4825 + £784 = £5609

Now try this calculation:
Using the table of income tax payments calculate how much income tax would have to be paid on the following taxable incomes:
(a) £6890   (h) £11 235   (c) £19 750   (d) £25 320

Answers on page 176.

Bonus payments and other additional payments

Additions to basic income such as bonuses and overtime make the calculations of income more complicated. The formulae for such calculations vary greatly, but the following examples give some standard methods.

Example
A firm pays a bonus of 15% to its employees when they exceed the

production targets set by the managers. Calculate how much bonus an employee would be paid if the standard wage was £225.30.

Solution

The bonus payment is 15% of £225.30
= 15/100 × £225.30 = £33.80

Example
The basic pay for a job is £2.80 per hour for a standard 40 hour week. If more than 40 hours are worked in a week and this is paid at 1.5 times the normal rate, calculate the gross earnings of a person who works 51.5 hours in a week.

Solution
Standard gross weekly wage is 40 hours multiplied by the rate for the job which is £2.80:

so 40 × £2.8 = £112.00

The number of hours of overtime = 51.5 − 40 = 11.5 hours
Rate of pay for overtime is 1.5 × £2.80 = £4.20 per hour
11.5 hours at £4.20 per hour = £48.30

Total gross weekly wage = £112.00 + £48.30 = £160.30

Now try these:
1. If the bonus payment in a cleaning company is 12.5% on the standard wage of £96.00 how much gross weekly income would be earned when the bonus is paid?
2. A local authority pays twice the normal rate when its employees work at weekends. If the standard weekly wage is £146.25 for a 37.5 hour week what would an employee's earnings be if he also worked 14 hours one weekend?

Answers on page 176.

Commission

Some salespeople work on a standard wage plus commission based on the value of the orders which they sell. The calculation of their gross earnings is similar to the previous example.

Example

A salesperson is given a basic salary of £3264 per year, and is paid a commission of 12.5% of the value of sales orders. If a person's sales orders for one month was £5232 calculate their gross earnings for that month.

Solution

standard gross monthly income = £3264/12 = £272
commission payment = 12.5/100 × £5232 = £654
gross monthly earnings = £272 + £654 = £926

A variation on commission is piece work or payment for actual work done; and, often, different levels of pay are made according to the level of production.

Example

The following table gives the payment in a production unit for each unit which is produced:

Units	Payment £
001–100	0.50
101–150	0.75
151–175	1.00
176+	1.50

Calculate the gross weekly earnings for a person who produces 183 units in a week.

Solution

Multiply each section by its respective rate until the total of 183 units is reached:

$$
\begin{array}{rl}
 & £ \\
100 \times 0.50 = & 50.00 \\
50 \times 0.75 = & 37.50 \\
25 \times 1.00 = & 25.00 \\
08 \times 1.50 = & 12.00 \\
\hline
\text{Total} \quad 183 \text{ units} & 124.50 \\
\hline
\end{array}
$$

Total gross wage = £124.50.

Now try these:

1. Using the table in the previous exercise calculate how much people would earn for each of the following levels of production:
    (a) 146   (b) 234   (c) 176   (d) 98
2. A young mother is an agent for a mail order firm and is paid 12.5% commission on all sales which she makes. If, during one year she sells goods valued at £5621 how much commission would she have earned?

Answers on page 177.

Increases in salary

Increases in salary resulting from negotiation are generally in percentage terms, although wage agreements can include flat rate increases.

Example
Calculate the new gross monthly salary of a person who earns £8520 per annum and is awarded a salary increase of 5.5% plus a flat rate increase of £250.

Solution
Generally the order of operations is to calculate the percentage increase, then to add this to the old salary and finally to add the flat rate increase:

$$\text{The percentage increase} = 5.5/100 \times £8520 = £468.60$$

New gross annual salary $= £8520 + £468.60 + £250 = £9238.60$

Now try this:
If your annual salary is £9864 and you are awarded a pay increase of 8% plus a flat rate increase of £400 what would your new salary be?

Answer on page 177.

Pensions

The amount of annual pension a person receives is generally based on the contributions made to the pension scheme. The contributions are a reflection of a person's earnings and length of service. Given these two variables people can calculate their pension.

Example
In a particular pension scheme the pension is based on the number of years of contribution divided by 80; this fraction is then multiplied by the average salary of the last three years of contributory service.

Using this formula calculate the annual pension of someone who earned £9220, £9900, and £10 512 in their final three years of service; and had contributed for 28.4 years.

Solution
The fraction which has been earned by the contributions is:

$$28.4/80 = 0.355$$

The average annual earnings of the final three years is:

$$(£9220 + £9900 + £10\,512)/3 = £29\,632/3 = £9877.33$$

Annual pension $= 0.355 \times £9877.33 = £3506.45$

Example
In the pension scheme described above a capital sum is given to the person when they retire. The sum is based on $\frac{3}{80}$ths of the average annual earnings for the final 3 years of service multiplied by each year of service.

Using this formula calculate the lump sum which would have been granted on the earnings given above.

Solution
The average annual earnings is £9877.33. To obtain the lump sum this average is multiplied by $\frac{3}{80}$ and then by the number of years service which is 28.4.

$$\frac{3}{80} \times £9877.33 = £370.40$$

The lump sum paid on retirement is £370.40 × 28.4 = £10 519.36

Now try these:

1. Assuming that you are a member of the pension scheme described in the example given above calculate the annual pension and lump sum if your earnings for the last three years were:

$$£11\,515 \qquad £12\,000 \qquad £12\,860$$

   and you had 31.30 years of contributory service.

2. Using the earnings given in the previous exercise, calculate the annual pension and lump sum if the contributory service had been 18.7 years.

Answers on page 177.

15

Gardening

Planning and the use of space

For many of us garden space is limited, and in order to make the best use of the space available we should plan its design. The design and requirements of each person's garden varies, and therefore gardening books can only give general advice on planning. The following example illustrates how such general information can be applied to a specific garden.

Example

The following table gives information about the production of Brussels sprouts.

Expected yield per plant	2 lbs
Minimum distance between plants	50 cm
Minimum distance between rows	25 cm
Approximate time between sowing and picking	30–40 weeks
Germination time	7–12 days
Picking starts	October

Calculate how much space, to the nearest square metre, is required to produce 10 lbs of Brussels sprouts per week for 16 weeks October to early January. Assume that 4 rows of plants would be the most convenient arrangement.

Solution
Each plant yields 2 lbs and to obtain 10 lbs for one week we would need 5 plants. For 16 weeks the minimum number of plants required is:

$$5 \times 16 = 80 \text{ plants}$$

The total length of row is 80 plants × distance between each plant = 80 × 50 cm = 40 metres.

Assuming that 4 rows would be convenient then the length of each row is:

$$40/4 = 10 \text{ metres}$$

Assuming that 25 cm is left on either side of each row then:

$$4 \text{ rows require 5 spaces of } 25\,\text{cm} = 5 \times 25\,\text{cm} = 125\,\text{cm}$$

The area required = 10 m × 1.25 m = 12.5 sq metres.

Now try the following:
The table gives information about the production of sprouting broccoli:

Expected yield per plant	1.5 lbs
Minimum distance between plants	60 cm
Minimum distance between rows	25 cm
Approximate time between sowing and picking	12 weeks
Germination time	7–12 days
Picking starts	August

You are required to calculate how much space to the nearest sq metre is required to produce 12 lbs per week for 13 weeks August to October. Allow for 4 rows to be planted, and the minimum distance of 25 cm should be allowed either side of each row.

Answer on page 177.

Timing in the garden

It is often useful to be able to draw a simple chart which shows the progress of a series of related tasks. This form of exercise is particularly useful in gardening where action at the right time is important.

Example
Using the information given above on Brussels sprouts construct a chart which highlights the timing of the tasks to be done.

Solution
We know, from the table, that the period between sowing and picking is 30 weeks. Since the picking starts in October the sowing must be 30 weeks earlier which is early March.

Using the information of page 123 we can construct the following chart:

```
          Jan/March   April/June   July/Sept   Oct/Dec      Jan
        |_____|_____|_____|_____|____|
Weeks 1      13           26          39          52    3

             9      to      25          39      to      3
       Week 9 ◄──────────────►          ◄──────────────►
              Sowing in 16 weeks         Picking in 16 weeks
```

Now try the following:
Using the information given above on sprouting broccoli construct a chart which highlights the timing of the production process.

Answer on page 177.

Area

A knowledge of how areas are calculated is important in gardening. Many garden designs include squares, rectangles and circles; and the application of seeds, fertilisers and feeds assumes that the user can calculate areas.

Example
You intend to make a lawn and shrub garden. The dimensions of the garden are 17 metres by 12 metres. Within that space you want 4

circular shrub areas each with a diameter of 2.5 metres. Calculate the area of the lawn and the area of each shrub circle.

Solution
The total area of the garden = 17 m × 12 m = 204 sq m.

Diameter of 1 shrub area = 2.5 metres.
Therefore the radius is 1.25 metres.

$$Area = Pi \times r \times r = 3.14 \times 1.25 \times 1.25$$
$$= 3.14 \times 1.56 = 4.89 \, sq \, m$$

For practical purposes the area for 1 shrubbery is 5 sq metres. Therefore the total area for shrubs is:

$$4 \times 5 = 20 \, sq \, m$$

The area of the lawn = 204 sq m − 20 sq m = 184 sq m.

Now try the following:
You are planning to make a herb and vegetable garden. The space available is 8 metres by 11.5 metres. It is intended that the herb garden should be circular with a diameter of 5 metres. Calculate how much space, to the nearest square metre, would be left for vegetables and paths.

Answer on page 177.

Mixing chemicals

When mixing fertilisers or pesticides for the garden we often have to calculate how much water is to be added, and this requires skill in converting imperial and metric measures, and in converting different types of measurement of volume.

Example
Your flower garden has been attacked by mildew, and the spray solution to control the mildew is applied at the following rates per square metre:

Rate 1 = 0.75 fluid oz of fungicide per 4.5 litres of water.
Rate 2 = 1.50 fluid oz of fungicide per 4.5 litres of water.

You estimate that the flower beds need 6 litres of rate 1, and the rose beds 2 litres of rate 2.

If 1 teaspoon = 0.2 fluid oz, how many teaspoons should be added to produce the required spray solution for the flower bed and the rose bed?

Solution
1. Flower beds need 6 litres of rate 1.
At rate 1 we know that 4.5 litres requires 0.75 fluid oz of fungicide:

so 1.0 litre requires 0.75/4.5 fluid oz = 0.17 fluid oz
6.0 litres requires 6 × 0.17 = 1.02 fluid oz

The number of teaspoons of the spray solution to be added to the 6 litres is:

1.02/0.2 = 5.1 teaspoons

2. Rose beds need 2 litres of rate 2.
At rate 2 we know that 4.5 litres requires 1.5 fluid oz of fungicide:

so 1.0 litre requires 1.5/4.5 fluid oz = 0.34 fluid oz
2.0 litres requires 2 × 0.34 = 0.68 fluid oz

The number of teaspoons of spray solution to be added to the 2 litres is:

0.68/0.2 = 3.4 teaspoons.

Now try the following:
You wish to control the weeds in your garden with a solution of weedkiller. The area to be sprayed is 25 sq metres. The contents of the weedkiller packet amount to 57 g, which is sufficient to treat 34 sq metres. Calculate how many grams of the weedkiller would be required to treat 25 sq metres.

Answer on page 177.

16

DIY

Fabric – measuring curtains

Most fabrics for curtains are sold in a standard width (panel) of 120 cm or 48 ins. In order to work out how many panels are needed the width of the window is multiplied by 2 to allow for gathering, and 10 cm is allowed for side hems.

Example

Assuming that the fabric you want to buy is sold in widths (panels) of 120 cm calculate how many panels of fabric would be required to make curtains for a window which is 1.8 m wide?

Solution

The number of panels required is the width of the window multiplied by 2 divided by the width of the fabric:

$$1.8\,m \times 2 = 3.6\,m = 360\,cm$$
$$360\,cm/120\,cm = 3$$

The number of panels needed is 3 and if 2 curtains are required then each curtain would need $3/2 = 1.5$ panels.

In order to calculate how much fabric should be bought to make curtains we also need to know the length of the window. To obtain the length of finished curtains measure from the level of the curtain hooks to the point you want the curtains to reach.

An allowance has to be made for hems at the top and at the bottom of the curtains. The bottom hem is normally 15 cm and the top hem is either 8 cm for a rod or 10 cm for rings.

Example
How much fabric should be purchased to provide curtains for a window which requires 3 panels and a finished length of 130 cm. The curtains are to be fixed to a rod.

Solution
The total length of fabric required is:

3 panels each with a length of 130 cm which is:
3 × 130 cm = 390 cm
Each of the 3 panels would have a bottom hem of 15 cm
3 × 15 cm = 45 cm
Each of the 3 panels would have a top hem of 8 cm for rods
3 × 8 cm = 24 cm

The total length required is 390 cm + 45 cm + 24 cm = 459 cm.

Now try these calculations:
Assuming that the fabric for curtains is sold in widths of 120 cm calculate the amount of fabric which would be required to make curtains for windows with the following dimensions:
1. Width = 2.2 m Finished length = 1.8 m Rings are to be used.
2. Width = 1.9 m Finished length = 1.4 m Rods are to be used.
3. Width = 3.2 m Finished length = 2.5 m Rings are to be used.

Answers on page 177.

Buying bricks

When estimating the number of bricks which are needed to build a wall, allowance should be made for the bed of mortar. The standard size of brick plus mortar is 225 mm (length) × 112.5 mm (width) × 75 mm (height). This nominal size can be used to estimate how many bricks should be ordered for a particular job.

Example

Estimate how many bricks are required to build a wall which is 1 brick wide, 5 m long and 30 cm high.

Solution

The number of bricks required for one course is the total length of the wall divided by 225mm. The length is 5 m or 5000 mm.

$$5000/225 = 22.2$$
Each course would require 22 bricks.

The number of courses is the total height of the wall divided by 75 mm. The height is 30 cm or 300 mm.

$$300/75 = 4$$
The wall would need 4 courses.

The estimated number of bricks needed is $22 \times 4 = 88$ bricks.

Now try these:

Estimate how many bricks would be required for the following jobs:
1. A wall which is 2 bricks wide, 4 m long and 450 cm high.
2. A wall which is 2 bricks wide, 6.2 m long and 2.2 m high. The wall includes a door which is 2 m high × 80 cm wide.

Answers on page 177.

Buying a carpet

Carpets are sold in standard widths of 2 m and 3 m. Carpet fitting, and estimating how much carpet is needed, is a skilled job; but very often, when looking at various alternative carpets, we want to estimate how much carpet we need to give us a guide to the total cost of fitting a carpet in a room. This is fairly straightforward when the room is a neat square or rectangle of 2 m or 3 m; when it is an awkward shape then the calculation is more complicated.

Example

Estimate the cost of fitting a carpet which costs £9.50 per sq m in a room with the following dimensions. The carpet is sold in widths of 3 m.

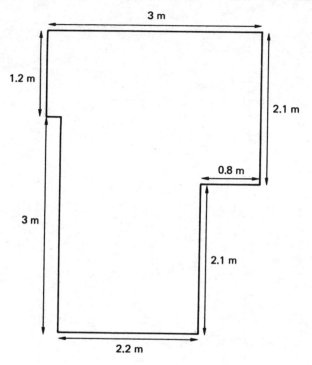

The maximum width required is 3 m
The maximum length is 1.2 m + 3 m = 4.2 m
The number of square metres which is needed is
3 m × 4.2 m = 12.6 sq m

The estimated cost of the carpet is 12.6 sq m × £9.50 = £119.70

Now try this:
Estimate the cost of fitting a carpet which costs £11.75 per sq metre
in a room with the following dimensions. The carpet is sold in widths
of 3 m.

Answer on page 178.

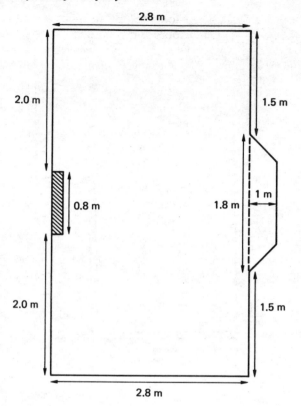

Buying paint

The volume of paint we have to buy for a job is determined by the manufacturer's estimate of the area which his brand of paint will cover plus an estimate of the area to be covered and the number of coats which will be applied.

Example

An emulsion paint, on average, covers 15 sq m per litre. Estimate how many litres are required to paint two coats on the walls of room which is 3.7 m wide, 4.2 m long and 2.4 m high. An allowance of 6 sq m should be made for doors and windows.

Solution

There are two walls of 3.7 m wide × 2.4 m high
The other two walls are 4.2 m wide × 2.4 m high

Area of wall to be painted is:
(2 × 3.7 m × 2.4 m) + (2 × 4.2 m × 2.4 m)
= 17.76 sq m + 20.16 sq m = 37.92 sq m

Deduct 6 sq m for doors, etc.
Area of the room to be painted is 37.92 − 6.0 = 31.92 sq m
This has to be painted twice
the total area to be painted is 31.92 sq m × 2 = 63.84 sq m

Volume of paint required is 63.84/15 = 4.25 litres.

Since this is an estimate 4 × 1 litre cans of paint should be purchased.

Now try these calculations:

Estimate how many litres of paint are required for the following jobs:

1. A room which is 2.9 m × 3.75 m and is 2.5 m high. The paint generally covers 9 sq m per litre and only 1 coat is required.
2. A room which is 3.0 m × 5.6 m and is 3.1 m high. Allow 10.5 sq m for doors and windows. The paint generally covers 11 sq m per litre and 2 coats would be needed.

Answers on page 178.

Buying wallpaper

Many charts are simply a convenient and quick method of estimating the requirements for a task. A good example is the charts which are widely available for estimating how much wallpaper is required to decorate a room.

Number of rolls of wallpaper required

Distance around the room	Height from skirting			
	2.2 m	2.5 m	2.8 m	3.1 m
10 m	4	5	5	5
12 m	5	6	6	7
15 m	6	7	7	8
17 m	7	8	9	10
19 m	8	9	10	11
22 m	9	10	11	12
24 m	11	13	14	16

No. of rolls

Example
Using the chart estimate how many rolls of wallpaper are required
for a room which is 4 m × 6.5 m and is 3 m high.

Solution
The distance around the room is 4 m + 6.5 m + 4 m + 6.5 m = 21 m.
The height of the room is 3 m. Reading the chart at the nearest
dimensions which are 22 m and a height of 3.1 m gives 12 rolls.
12 rolls should therefore be purchased.

Now try this:
Using the chart estimate how many rolls of wallpaper are required
for a rooms of the following dimensions:
1. 6.3 m wide, 5 m long and 2.8 m high
2. 4 m wide, 3.25 m long and 2.2 m high
3. 5.5 m wide, 4.7 m long and 2.7 m high.

Answers on page 178.

Ordering concrete

The amount of concrete required for a job depends on the area to be
covered and the thickness of concrete which is needed. The following
chart gives an estimate of the amounts of cement, sand and
aggregate which make up two standard mixes of cement.

Chart showing how much concrete to order

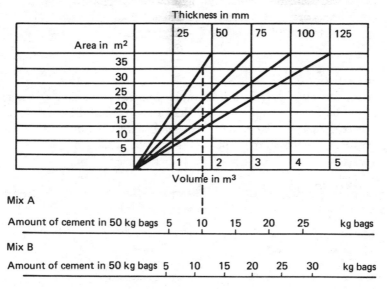

Mix A

Amount of cement in 50 kg bags   5   10   15   20   25   kg bags

Mix B

Amount of cement in 50 kg bags   5   10   15   20   25   30   kg bags

Mix A (1:2.5:4)   1 bag cement
              $2\frac{1}{2}$ bags sand
              4 bags aggregate

Mix B (1:2:3)   1 bag cement
             2 bags sand
             3 bags aggregate

Example

Using the chart calculate how much cement, sand and aggregate to order to the nearest 50 kg bag for Mix A if the area of concrete required is 7 m × 5 m and the thickness is 50 mm.

Solution

$$\text{Area} = 7\,\text{m} \times 5\,\text{m} = 35\,\text{sq m}$$

Find the intersection of the area on the horizontal line and the thickness. Reading the chart we see that an area of 35 sq m and thickness of 50 mm gives a volume of less than 2 cubic metres. The estimate of the ingredients required is:

Cement: 10 × 50 kg bags

Once we have the number of bags of cement we know that for mix A the proportions to cement:sand:aggregate are 1:2.5:4.

Sand: $(2.5 \times 10 \times 50 \, kg) = 25 \times 50 \, kg$ bags

Aggregate: $(4 \times 10 \times 50 \, kg) = 40 \times 50 \, kg$ bags

Now try this:

Using the chart calculate how much cement, sand and aggregate should be ordered to the nearest 50 kg bag for the following:

1. Mix B.   Area = 20 sq m   Thickness =   75 mm
2. Mix A.   Area = 7 m × 5 m   Thickness = 100 mm
3. Mix B.   Area = 3.2 m × 4.8 m   Thickness = 50 mm

Answers on page 178.

17

Housing

Mortgage

Most people, when buying a house, borrow some of the money on a mortgage. Although there are many types of mortgage arrangements the repayment of the loan generally comprises two main elements: the repayment of part of the outstanding loan and repayment of interest on the loan.

There are many methods of repaying a mortgage, the following example concentrates on the calculation of interest on the loan which is outstanding.

Example

You owe £18 000 on a mortgage loan and the annual interest charged on the loan is 11.5%. Calculate the amount of annual interest due.

Solution

11.5% of £18 000 is 11.5/100 × £18 000 = £2070

Now try these calculations:
Calculate the amount of annual interest which would be due on the following mortgage arrangements:
1. A loan of £14 000 at 12% per annum.
2. A loan of £32 500 at 11.75% per annum.
3. A loan of £25 650 at 13.25% per annum.

Answers on page 178.

Tax relief

Tax payers can generally claim tax relief on the payment of interest on mortgage loans. Currently the relief is at the standard rate of income tax for loans up to £30 000.

Example

You owe £24 500 on a mortgage and the annual rate of interest is 12.5%. (i) Calculate the gross interest which is due. (ii) Assuming that you can claim tax relief on the interest which is due and that the standard rate of taxation is 25%, how much actual interest would be paid?

Solution

(i) The gross interest which is due is 12.5% of £24 500 is:

$$\tfrac{12.5}{100} \times £24\,500 = £3062.50$$

(ii) Since tax relief of 25% of £3 062.50 is allowed then the actual amount to be paid is:

$$100\% - 25\% = 75\% \text{ of } £3062.50$$
$$= \tfrac{75}{100} \times £3062.50 = 2296.88$$

Now try this:
Assuming that you can claim tax relief on the interest paid on loans of up to £30 000 and that the standard rate of tax is 25%, calculate how much actual interest would be paid on the following mortgage loans:
(a) A loan of £19 600 at 13%
(b) A loan of £34 000 at 11.5%
(c) A loan of £62 000 at 12.25%

Answers on page 178.

Buying and selling houses

Moving house can be an expensive operation, and it is not always easy to estimate what the costs will be. The following lists give estimates of the most common costs incurred when moving house.

Example
Estimated costs of selling a house:

>   Estate Agent's fee is 1.5% of selling price
>   Solicitor's fee is 1.0% of selling price
>   Removal costs are 0.8% of the selling price

Estimated costs of buying a house:

>   Solicitor's fee is 1.2% of buying price
>   Survey fees are 0.5% of the buying price

Stamp duty is charged on the total price, and is currently levied at the following rates (£):

Up to £25 000	0.0% of buying price
£25 001 to £30 000	0.5% of buying price
£30 001 to £35 000	1.0% of buying price
£35 001 to £40 000	1.5% of buying price
over £40 000	2.0% of buying price

Mr D Crab sells a house for £56 000 and buys another for £69 000. Using the estimates given above calculate the cost of buying and selling.

Solution
Selling costs:

Estate Agent	$\frac{1.5}{100} \times £56\,000 =$	£840.00
Solicitor	$\frac{1.0}{100} \times £56\,000 =$	£560.00
Removals	$\frac{0.8}{100} \times £56\,000 =$	£448.00
	Total selling costs	£1848.00

Buying costs:

Solicitor	$\frac{1.2}{100} \times £69\,000 =$	£828.00

$$\text{Surveyor} \qquad \tfrac{0.5}{100} \times \text{£69\,000} = \quad \text{£345.00}$$
$$\text{Stamp duty} \qquad \tfrac{2.0}{100} \times \text{£69\,000} = \text{£1380.00}$$

$$\text{Total buying costs} \qquad \text{£2553.00}$$

Total costs = £1848.00 + £2553.00 = £4401.00

Now try the following:

Mrs L Fish sells a house for £85 000 and buys a bungalow for £120 000. Using the estimated costs given in the previous example calculate the costs which Mrs Fish incurs in buying and selling.

Answer on page 178.

18

Borrowing

Interest rates

When money is borrowed, for example on a bank loan, interest has to be paid. Interest is the cost of borrowing. There are various factors which determine the rate of interest which we would have to pay on a loan. The most important factor is the general level of bank interest rates. In addition conditions such as the extent of our security and the reasons for borrowing the money are taken into account.

Most loans are charged at only one specific rate of interest; and it is the rate and the timing of the repayments of the loan which determine how much the loan will cost. So, if we wish to compare the cost of various methods of borrowing money, it is important not only to know the actual rate of interest but also to understand the different systems of repayment.

Flat rate repayments

When we repay a loan there are two parts to the repayment. Firstly we must pay back the total amount of money which has been borrowed, and secondly we must pay interest on the money. In the following example we shall assume that the total amount of money which has been borrowed is paid back, in full, at the end of the loan period.

Example

A rate of interest of 12% per annum is charged on a loan of £1600. If the loan is outstanding for 3 years how much interest will be due each year, and what is the total cost of the loan?

Solution

We assume that we shall repay £1600 at the end of three years and the cost of the loan is the interest paid. The annual rate of interest is 12% of £1600:

$$(\tfrac{12}{100}) \times £1600 = £192$$

The loan is for 3 years.
The total cost of the loan is $3 \times £192 = £576$.

Example

What would the cost of the above loan be if it were to be repaid after 15 months?

Solution

12 months costs £192 and therefore 15 months costs:

$$(\tfrac{15}{12}) \times £192 = £240$$

We can see from these two examples that the cost of a loan is determined by the period of time the loan is outstanding as well as by the actual rate of interest.

Now try these calculations:

1. You take out a loan of £5600 for ten years at a rate of interest of 17%, assuming that you repay the loan in full at the end of ten years how much would the loan cost?
2. If, in the previous exercise, you repaid the loan in 7 years calculate how much the loan would cost.

Answers on page 178.

True rate of interest

When the amount borrowed is not repaid until the end of a period, as in the previous examples, then the flat rate of interest is also the true rate. But if the loan is repaid in instalments then the flat rate of interest is not the true rate.

Example

A rate of interest of 15% is charged on a 3 year loan of £2400. The amount of interest to be paid on the loan is calculated to be $3 \times (\frac{15}{100}) \times £2400 = £1080$. The total amount of money owed at the end of three years is the interest which is owed plus the amount borrowed:

$$£1080 + £2400 = £3480$$

If the borrower repays the loan in six equal instalments of £580 over 6 months can it be shown that the flat rate of interest of 15% is not the true rate of interest?

Solution

If we assume that each repayment of £580 includes £400 as repayment of the loan – this is calculated by dividing the loan of £2400 by 6 payments – then the remainder of the instalment of £180 is payment of interest. The actual amount of loan which is outstanding at the end of each instalment period is given in the following table.

Instalment period	Amount repaid £	Outstanding balance £
Opening balance		2400
1	400	2000
2	400	1600
3	400	1200
4	400	800
5	400	400
6	400	000

We can use a formula to calculate the true rate of interest.

The formula is: $R = (I \times 100)/(P \times Y)$ where:

> R = true rate of interest
> I = the amount of interest paid
> P = the amount of the loan outstanding
> Y = repayment or instalment period (in years)

In the example given above we know that the amount of interest paid each month (I) = £180 and that the repayment period (Y) is 6 months or 0.5 years.

We also know the different amounts of the loan outstanding (P) at the end of each period of instalment. Using this information we can use the formula to construct the following table.

Instalment	Formula	True rate %
1	$(180 \times 100)/(2400 \times 0.5) = 15.00$	
2	$(180 \times 100)/(2000 \times 0.5) = 18.00$	
3	$(180 \times 100)/(1600 \times 0.5) = 22.50$	
4	$(180 \times 100)/(1200 \times 0.5) = 30.00$	
5	$(180 \times 100)/(\ 800 \times 0.5) = 45.00$	
6	$(180 \times 100)/(\ 400 \times 0.5) = 90.00$	
Total		220.50

We can see from the table that the flat rate of interest of 15% applies only to the first instalment period, after that the true rate increases with each instalment period. This is because the amount outstanding on the loan is the actual money borrowed, but the flat rate of 15% assumes that no repayment of the loan has taken place.

In this example the Annual Percentage rate (APR) is calculated by finding the total of the rates paid and dividing by 6. This gives:

$$220.50/6 = 36.75\%$$

Now try this:
You wish to borrow £5000 to buy a car. Calculate the APR for each loan:

Loan 1:   Amount borrowed = £5000
            Period of loan = 5 years
            Rate of interest = 17% per annum on £5000
            Repayments in equal instalments every 6 months.

Loan 2:   Amount borrowed = £5000
            Period of loan = 5 years
            Rate of interest = 20% per annum on £5000
            Repayments in 5 equal annual instalments.

Answers on page 178.

Credit and hire purchase

There are a variety of organisations which are in business to lend money: commercial banks, building societies, finance companies and so on. The terms for loans offered by this range of organisations varies widely and it is useful to be able to calculate the different costs of borrowing.

Example

You want to buy a car for £10 900 and you can part exchange your existing car for £4000. Calculate the monthly repayment rates of the following loan assuming that you make a deposit of £4000.

The car dealer can arrange a loan at an interest rate of 9.5% per annum over two years providing that a deposit of at least $\frac{1}{3}$ is made. The interest rate of 9.5% is to be paid over 2 years on the full amount which is borrowed.

Solution

The deposit which is required is $\frac{1}{3}$ of £10 900
$$= £3633$$

The amount to be repaid over 2 years is £10 900 − £3633
$$= £7267$$
The amount to be repaid in 1 year is £7267/2
$$= £3633.50$$
The cost of annual interest is 9.5% of £7267
$$= £690.37$$
The amount to be repaid in one year is
$$£3633.50 + £690.37 = £4323.87$$

The monthly repayments are £4323.87/12 = £360.32

Now try these calculations:

1. You want to buy a new kitchen and the total cost of installation is £6900. A finance company is prepared to lend you £6900 over

three years. The annual rate of interest on the loan is 11% and this has to be paid on the total amount borrowed over the three years. Calculate how much the loan would cost each month.

2. Using the information in the previous problem calculate how much the loan would cost each month if it was £5400 at 9.75%.

Answers on page 178.

19

Banking

There are two main types of bank account: deposit account and current account.

A deposit account is not designed for everyday transactions: the banks use the money placed in deposit accounts to lend to borrowers. Banks therefore require notice of withdrawals from deposit accounts, and, because they are using the money, interest is paid on the account.

On the other hand a current account is a service offered by banks to enable people to pay bills by cheque, by standing order and by direct debit. A charge is generally made for this service.

Calculation of current account charges

Charges for their services vary from bank to bank, but they do have two main parts. Firstly, charges are not generally made if there is a minimum amount of money in the account. Secondly, if the current account balance is less than the sum required then a charge is made for cheques issued, statements, standing orders and direct debits. The charging period is generally a quarter.

The following table is typical of the charges made by the leading commercial banks:

	Minimum Balance £	Cost of transaction £
Lloyds	100.00	0.20
Midland	100.00	0.20
Barclays	50.00	0.175

Nat West	50.00	0.20
Co-op	00.01	0.20

Example

The following bank statement shows the transactions of Mr and Mrs Gibbs during a single month. Calculate their bank charges assuming that:

(i) no charges are made if the balance in the account exceeds £100;
(ii) the cost of each debit transaction is 17.5p;
(iii) withdrawals from the cash dispensing machine cost 12p.

HIGH STREET BANK
Statement of Account. Gibbs E and G   April.

Date	Particulars	Payments £	Receipts £	Balance £
1	Opening balance			425.22
9	Action Aid S/O	7.50		417.72
10	Loan 0253464	305.89		111.83
11	374925	27.00		84.83
12	Cross Ass'n D/D	15.00		69.83
13	Brent C/P 3	50.00		19.83
16	374924	11.45		8.38
17	374929	5.98		2.40
18	Sundry Credit		174.21	176.61
19	Derby C/P 3	70.00		106.61
24	Yorks Gen D/D	13.47		93.14
25	374935	19.23		73.91
26	Avon CC BGC		804.06	877.97
30	Closing Balance			877.97

Solution

Number of cash machine transactions (CPs) is 2

$$\text{cost} = 2 \times 12p = 24p$$

Number of payments when balance is below £100 is 7

$$\text{cost} = 7 \times 17.5p = £1.23$$

Total bank charges $= £0.24 + £1.23 = £1.47$

Now try this:

The following bank statement shows the transactions of Mrs King during a single month. Calculate her bank charges assuming that:
(i) no charges are made if the balance in the account exceeds £50;
(ii) the cost of each debit transaction is 20p;
(iii) withdrawals from the cash dispensing machine cost 12p.

HIGH STREET BANK

Statement of Account. King F H   June

Date	Particulars	Payments £	Receipts £	Balance £
1	Opening Bal.			612.58
9	NSPCC S/O	10.00		602.58
10	Loan 459829	125.90		476.68
11	492820	372.89		103.79
12	Fox Guild D/D	25.00		78.79
13	Bristol C/P 3	50.00		28.79
16	492827	16.78		12.01
17	492830	28.90		−16.89
18	Sundry Credit		120.00	103.11
19	Bristol C/P 3	50.00		53.11
24	Bath Ins. D/D	12.54		40.57
25	492831	22.50		18.07
26	Avon CC BGC		543.90	561.97
30	Closing Balance			561.97

Answer on page 178.

Credit cards

Generally, in order to pay bills by cheque, you must have a current account in a commercial bank. A method of payment which is becoming more common is payment by credit card such as Access or Barclaycard.

Access and Barclaycard charge interest on a monthly basis on the average daily debt. The interest is charged on the balance from the previous monthly statement.

Example

Using the following information calculate the interest due and the balance of Mrs B Sharp's credit card statement. The date of the previous statement was 4th March, and interest is charged at 2% per month.

B Sharp 4926-435-210-721   Statement date: 4th April.

	£
Balance from previous statement	282.50
24th March Payment received	60.00
New Balance before interest charges	222.50

Solution

First calculate the average debt for the period of the statement, which is 4th March to 4th April, or 31 days.

The number of days the debt of £282.50 outstanding 4th March to 24th March = 20 days.
Multiply the debt by the number of days:

$$20 \times £282.50 = £5650$$

The number of days remaining is:

$$24th\ March\ to\ 4th\ April = 11\ days$$

Multiply the 11 days by the remaining debt:
$$11 \times £222.50 = £2447.50$$

Add the two sums £5650 + £2447.50 = £8097.50

The sum of £8097.50 represents the total debt for one month. The average numbers of days in a month is:

$$365/12 = 30.42$$

To find the average daily debt divide the sum by the average number of days per month.
Average daily debt for the month of March is:

$$£8097.50 \div 30.42 = £266.19$$

The interest charged is 2% of £266.19

$$\tfrac{2}{100} \times 266.19 = £5.32$$

The balance brought forward on 4th April is:
$$£222.50 + £5.32 = £227.82$$

Now try this:
Calculate the interest due and the balance of Mrs Star's credit card statement. The date of the previous statement was 9th July and interest is charged at 2.5%.

K Star. 429-123-456-789   Statement date 9th August

	£
Balance from previous statement	1023.65
4th August Payment received	380.00
New balance before interest charges	643.65

Answer on page 178.

Bank reconciliation

When a bank statement is issued it might not show all of the transactions which have taken place during the period of the statement. Discrepancies occur because cheques which have been used to pay bills have not been presented to the bank. In order to find out the true position such cheques have to be taken into account. This type of calculation is known as bank reconciliation.

Example
During June, Mr Wall issued 12 cheques which were numbered from 052330 to 052341. In addition, on the 27th June, Mr Wall made a payment of £25.00 into his current account. Using this information and the following statement determine Mr Wall's true current account balance.

HIGH STREET BANK

Mr. S Wall   Bank Statement.   June   £

Date	Particulars	Payment	Receipts	Balance
	Opening Balance			432.91
8	052330	60.00		372.91
9	Lloyds Bowmaker D/D	32.78		340.13
12	052332	71.25		268.88
14	052333	25.00		243.88
16	052336	17.50		226.38
16	Loan 0254641	121.89		104.49
19	052339	5.63		98.86
20	052337	36.22		62.64
22	Sundry Credit		47.91	110.55
23	052331	89.20		21.35
26	052340	15.00		6.35
28	Sun Alliance D/D	51.05		−44.70
30	Closing Balance			−44.70

The following is the value of the cheques not presented:

Cheque No.	£
052334	23.21
052335	50.00
052338	42.17
052341	11.20
Total	126.58

Answer on page 179.

Solution
The bills which have been paid but have not been included in the bank statement must be deducted from the closing balance. Similarly the money paid into the bank but not shown on the statement must be added to obtain the new balance.

	£
28th June Balance	−44.70
Less cheques not presented	−126.58
	−171.28
Add sundry credit	+25.00
Reconciled Balance	−146.28

Mr Wall's overdraft is £146.28

Now try this:
During July Mr Wall paid bills with 11 cheques which were numbered from 052342 to 052352. In addition, on the 25th July, Mr Wall received a payment of £35.87 into his current account. Using this information and the following statement determine Mr Wall's true current account balance. The value of the unpresented cheques is given on the following page.

HIGH STREET BANK
Mr. S Wall   Bank Statement.   July   £

Date	Particulars	Payment	Receipts	Balance
	Opening Balance			−146.28
2	052334	23.21		−169.49
4	Brown & Son BGC		698.25	528.76
5	052338	42.17		486.59
8	052343	40.00		446.59
9	Lloyds Bowmaker D/D	32.78		413.81
12	052335	50.00		363.81
14	052345	45.90		317.91
16	052341	11.20		306.71
16	Loan 0254641	121.89		184.82
19	052346	14.32		170.50
20	052347	32.89		137.61
21	Sundry Credit		56.20	193.81
23	052350	43.90		149.91
26	052349	15.56		134.35
28	Sun Alliance D/D	51.05		83.30
31	Closing Balance			83.30

The following is the value of the cheques not presented:

	£
052342	14.52
052344	20.00
052348	31.12
052351	62.14
052352	15.50
Total	143.28

Answers on page 179.

20

Budgeting

Budgeting uses several arithmetic skills, including estimating. A budget is a plan which is used to control spending; and if it is to be successful then estimates about future spending must be realistic.

The simplest approach to a budget is to have a look at how money has been spent during the previous year, and to assume that the pattern will continue in the coming year. For instance, if we spent £128.82 on telephone bills in one year then it would be reasonable to assume that we shall spend a similar amount in the following year. Because a budget is an estimate it would be unrealistic to assume that our 'informed guess' would be exact, and therefore an estimate of £130 would be acceptable.

Furthermore, if we assumed that telephone charges might increase during the following year, then we could add our estimate of the increase to last year's expenditure.

For instance, if our estimate of the increase was 10% then the budgeted figure is:

$$£130 + 10\% \text{ of } £130 = £130 + \tfrac{10}{100} \times £130$$
$$= £130 + £13 = £143$$

Fixed spending

Quite a lot of our spending is fixed: it cannot be avoided. For example, most people have to pay rent or repay a mortgage, many of us will be paying the community charge, everyone has to buy food.

So, in any budget, money must be set aside to pay for the fixed part of our spending.

Example

The following list gives the actual amount spent during one year, and the anticipated percentage increase in expenditure on each item.

	£	% increase
Rent	1832.75	(0%)
Community charge	399.86	(15%)
Insurance	365.24	(20%)

Prepare a budget based on this information.

Solution

Item	Actual spent	Rounded figure	Expected change %	Budget £
Rent	1832.75	1833	0	1833
Community charge	399.86	400	15	460 (£400+15% of £400)
Insurance	365.24	365	20	438 (£365+20% of £365)

NB: Since a budget is an approximation we can round off the actual figures before calculating the budgeted values.

Now try these calculations:

The following table gives the actual amount spent by Mr Moor during one year, and also shows his estimate of the percentage increase in expenditure during the coming year.

	Actual spent £	Increase %
Rent	1577.35	10
Community charge	257.84	15
Insurance	312.50	5

1. Prepare a budget for all three items.
2. Assume that you expect insurance premiums to increase by 8% rather than by 5% and calculate the difference this would make to the budget.

Answers on page 179.

Semi-fixed spending

Some items of expenditure, such as electricity or food, are partly fixed; but, unlike rent or mortgage, spending on these items can be controlled. For example, we can buy cheaper food and we can use less electricity. Even so we still have to eat food and, during winter, we have to keep warm.

The following list of annual expenditure shows some of those items which, to some extent, we can control although they are partly fixed.

	Cost £
Electricity	342.68
Telephone	235.23
Food	1872.56
Motor car	873.60

The arithmetic skill required to prepare a budget for these items is slightly more complicated than for those items which are wholly fixed. The cost of food, for instance, is affected by inflation and we would expect prices to increase; but, because we have some control over these items, we can plan to make savings and our estimate of how much we would want to spend can be more accurate.

Example

Using the list given above prepare a budget in which it is assumed that there will be a general increase in prices of 15%, but that savings of $\frac{1}{3}$ on each item could be made.

Solution

		Budget before Savings £
Electricity	£343 + 15% of £343	= 394
Telephone	£235 + 15% of £235	= 270
Food	£1873 + 15% of £1873	= 2154
Motor car	£874 + 15% of £874	= 1005

		Budget after savings £
Electricity	394 − (1/5 × 394) =	315
Telephone	270 − (1/5 × 270) =	216
Food	2154 − (1/5 × 2154) =	1723
Motor car	1005 − (1/5 × 1005) =	804

Now try this:

Using the following information prepare an annual budget for spending on electricity, telephone, food and motor car.

Annual Expenditure £	
Electricity	190.72
Telephone	132.49
Food	843.21
Motor car	649.82

Base your calculations on the assumption that the general increase in prices will be 7% and you could manage a saving of 5%.

Answers on page 179.

Quarterly payments

Many items which are partly fixed have to be paid each quarter, that is, every three months; in order to include them in a monthly budget we must divide the actual bill by 3. When quarterly payments are included in an annual budget the amount paid must be multiplied by 4.

Example

The following list shows the quarterly payments made on three items:

	£
Electricity	64.30
Telephone	85.67
Gas	56.73

Show the value of these items if they were included in a monthly and annual budget.

Solution

	Quarterly Payment £	Monthly Budget £
Electricity	64.30	$(64.30)/3 = 21.43$
Telephone	85.67	$(85.67)/3 = 28.56$
Gas	56.73	$(56.73)/3 = 18.91$

	Quarterly Payment £	Annual Budget £
Electricity	64.30	$4 \times 64.30 = 257$
Telephone	85.67	$4 \times 85.67 = 343$
Gas	56.73	$4 \times 56.73 = 227$

It is important to remember that budgets are estimates of future spending not exact figures, and so the calculations can be rounded up or down. In this example, although the estimated total spending has been made on the actual payment, the result has been rounded to the nearest pound.

Now try this calculation:

Using the method shown in the above example, calculate the monthly budget figure and the annual budget figure for the following quarterly payments:

	£
Electricity	74.23
Telephone	39.16
Gas	54.98

Answers on page 179.

Variable spending

The budget figures for items of fixed and partly-fixed spending are
relatively easy to calculate. Firstly, records of previous payments are
usually available; and secondly, the amount spent each month or
each quarter tends to remain the same. On the other hand, much of
our spending on such things as clothes or entertainment can vary
from month to month, and the amount which is spent on such items
is not always recorded.

If a budget is to be successful, then a record of the amount spent
on various items must be kept. In the following examples we shall
assume that expenditure has been recorded.

Example

The following shows the amount spent on leisure and holidays by
Miss Keen during one year January to December:

Jan	Feb	Mar	Apr	May	June £
15	19	12	62	23	18

Jul	Aug	Sep	Oct	Nov	Dec £
14	217	11	16	16	69

Using this expenditure show how much you think Miss Keen should
allow for leisure and holidays in a monthly budget.

Solution

In order to prepare the monthly budget on leisure and holidays, we
must calculate the amount which is spent on average each month,
and use this average as the budget figure for the following year.

The average monthly spending is calculated by dividing the total
annual spending by 12 (months). We find that the total spent by
Miss Keen was:

$$£15 + £19 + £12 + £62 + £23 + £18 + £14 + £217 + £11 + £16 + £16$$
$$+ £69$$
$$= £492$$

Miss Keen's average monthly spending $= 492/12 = £41$.

Now try this:

The following is a record of expenditure on entertainment by Mr Jones during one year January to December:

Jan	Feb	Mar	Apr	May	June £
24	28	28	25	36	52

Jul	Aug	Sep	Oct	Nov	Dec £
30	56	22	28	24	41

1. Calculate the monthly average spending by Mr Jones to the nearest whole number.
2. Assume that the June figure includes £25 whiich is the cost of a dinner dance which Mr Jones will not attend during the coming year. If this is eliminated what difference would it make to the monthly budget estimate?
3. Calculate the monthly budget estimate assuming there will be a 5% increase in prices.
4. Mr Jones decides that he will have a similar monthly spending pattern but that he will not spend more than £35 in any one month on entertainment. Reduce those months which exceed £35 to £35 and calculate the revised monthly budget.
5. Alternatively Mr Jones decides to increase his spending on entertainment by 12%; what effect would this have on the monthly budget?

Answers on page 179.

Monthly budget

If you have practised applying the four rules, calculating percentages and averages; and become used to estimating and rounding, then you are in a position to work out a monthly budget, and to estimate how much should be spent and how much can be saved each month.

Example

The following example shows the monthly budget calculation of a family of four. The layout in this example is a suggested layout; other layouts could be used:

		£	£
Net monthly income:			
Husband		509.13	
Wife		390.37	
	Total	899.50	899.50
Deduct fixed monthly expenses:			
Mortgage		262.25	
Community charge		24.50	
Insurance		38.67	
Hire Purchase		42.86	
	Total	368.28	368.28
			531.22
Deduct monthly estimate of annual and quarterly payments:			
Gas		26.00	
Electricity		28.00	
Telephone		17.00	
Motor car		51.00	
Insurance		14.00	
	Total	136.00	136.00
			395.22
Deduct monthly estimate of variable expenses:			
Food		220.00	
Milk		24.00	
Clothes		40.00	
Holidays		50.00	
Leisure		35.00	
	Total	369.00	369.00
Estimated monthly saving			£26.22

Now try these calculations:
1. Using the budget information in the previous example,
    (i) Calculate what the savings would be for that month if:
        (a) An unexpected bill for £64.32 was received?
        (b) The quarterly gas bill was actually £82.50?
        (c) The Community charge was increased by 20%?
    (ii) What is the estimated annual cost of the telephone?
    (iii) What percentage of total income are the variable expenses?
2. Prepare a similar monthly budget based on your own actual income and anticipated expenditure.

Answers on page 179.

Answers

Answers to Section One

Using numbers

Page 6
1. (a) 625   (b) 10 802   (c) 52.264   (d) 135.736   (e) 34.224
2. (a) 87.541   (b) 15.9497
3. £10.80   4. 13 042

Page 8
1. (a) 1324   (b) 1869   (c) 102⁹
2. 26.873   3. 1.501   4. 133.46   5. 7.469   6. 57   7. 825

Page 9
1. 9.032   2. 0.636   3. 5.353

Page 11
1. (a) 18   (b) 36   (c) 40   (d) 7   (e) 54   (f) 72
2. £12.00   3. 28   4. 6

Page 12
(a) 21   (b) 30.1   (c)   0.4   (d) 432   (e) 7390   (f) 0.61
(g) 6530   (h) 12 001   (i) 30   (j) 1053.4   (k) 93 000   (l) 70

Page 14
1. (a) 7   (b) 3   (c) 4   (d) 6   (e) 7   (f) 9   (g) 7   (h) 8
(i) 8   (j) 6
2. 12   3. 4

Page 15
1. (a) 305   (b) 674   (c) 500
2. (a) 248   (b) 740   (c) 406
3. (a) 235   (b) 700   (c) 36 750   (d) 4903   (e) 10   (f) 38
4. (a) 0.51   (b) 7.36   (c) 0.004   (d) 3.16   (e) 0.8749

(f) 0.00601   (g) 0.9382   (h) 0.00001   (i) 0.00016
(j) 3.974205   (k) 0.0009037

Page 16
(a) 13   (b) 3   (c) 11   (d) 1   (e) 13   (f) 1   (g) 100
(h) 8   (i) 0

Page 18
1. (a) 16.4   (b) 0.1   (c) 3.0   (d) 12.6   (e) 9.1
2. (a) 16.42   (b) 0.07   (c) 3.04   (d) 12.60   (e) 9.06

Using a calculator

Page 22
(a) 30.321   (b) 6.818   (c) 8.005   (d) 9632.77   (e) 27.66378
(f) 1.8415233

Page 24
(a) 10.7   (b) 46.4   (c) 15.1   (d) –11.8   (e) 14

Page 25
(a) £3.36   (b) £9.20   (c) £1.44   (d) 89p   (e) £10.40
(f) 4.051   or   4050 ml   (g) 1.55 kg   or   1550 g

Fractions

Page 27
(a) $\frac{1}{4}$ (b) $\frac{1}{6}$ (c) $\frac{3}{4}$ (d) $\frac{5}{8}$

Page 28–29
(a) $\frac{1}{4} = \frac{2}{8} = \frac{3}{12} = \frac{4}{16}$
(b) $\frac{3}{5} = \frac{6}{10} = \frac{18}{30} = \frac{15}{25}$
(c) $\frac{2}{14} = \frac{1}{7} = \frac{3}{21} = \frac{8}{56}$
(d) $\frac{10}{12} = \frac{5}{6} = \frac{15}{18} = \frac{40}{48}$

Page 31
1. (a) $2\frac{1}{2}$ (b) $1\frac{5}{7}$ (c) $1\frac{1}{3}$ (d) $5\frac{1}{7}$
2. (a) $\frac{37}{7}$ (b) $\frac{13}{8}$ (c) $\frac{39}{4}$ (d) $\frac{30}{17}$

Page 31
(a) $29\frac{2}{5}$ (b) $261\frac{3}{10}$ (c) $\frac{9}{10}$ (d) $3\frac{61}{100}$ (e) $40\frac{3}{4}$ (f) $9\frac{1}{25}$

(g) $12\frac{181}{500}$ (h) $7\frac{351}{500}$ (i) $38\frac{117}{1000}$ (j) $\frac{9}{5000}$

Page 32
(a) 1.3 (b) 21.7 (c) 6.39 (d) 13.07 (e) 0.53 (f) 5.021
(g) 8.984 (h) 0.1284

Page 33
(a) 0.75 (b) 0.4 (c) 0.5 (d) 0.267 (e) 0.375 (f) 0.3125
(g) 4.25 (h) 7.2 (i) 13.35 (j) 9.625 (k) 3.92

Page 33
(a) $1\frac{1}{8}$ (b) $1\frac{9}{11}$ (c) $3\frac{3}{5}$ (d) 2

Page 34
(a) $\frac{1}{3}$ (b) $\frac{2}{7}$ (c) $\frac{4}{15}$ (d) $\frac{1}{3}$

Page 35
(a) $\frac{11}{12}$ (b) $1\frac{7}{30}$ (c) $1\frac{5}{28}$ (d) $3\frac{7}{12}$ (e) $4\frac{3}{40}$ (f) $7\frac{13}{24}$ (g) $\frac{1}{3}$

(h) $1\frac{13}{28}$ (i) $3\frac{17}{42}$

Page 36
(a) $1\frac{1}{2}$ (b) $4\frac{5}{6}$ (c) $\frac{5}{8}$

Page 37
(a) 8 (b) 4 (c) $\frac{1}{3}$ (d) $3\frac{18}{23}$ (e) $2\frac{1}{8}$ (f) $1\frac{1}{2}$

Measuring quantities

Page 42–43
1. 2300 ml 2. 2 3. 4.6 kg 4. 1.411 5. 690 g 6. 3.9 m
7. 2.73 kg 8. 8 9. 2600 mm, yes.

Ratios and proportions

Page 45
(a) 2:9 (b) 2:1 (c) 1:5 (d) 7:2 (e) 7:2

Page 45
(a) 9:25 (b) 3:10 (c) 3:20

Page 46
(a) A = £600, B = £120, C = £360, D = £720
(b) Ammonia = 364 g, Superphosphate = 727 g, Bonemeal = 727 g, Potash = 182 g

Page 46–47
(a) £1.40   (b) £24.00   (c) £4.05   (d) £7.38

Page 47
(a) 24p   (b) 32p   (c) £2.50   (d) 30p

Page 47–48
(a) 40 mins   (b) 24 mins

Page 48–49
(a) 2.5 hours   (b) 10.5 hours

Page 49
(a) 1 hr 36 mins   (b) 2 hrs 5 mins

Percentages

Page 51
(a) $\frac{1}{20}$ (b) $\frac{1}{5}$ (c) $\frac{3}{20}$ (d) $\frac{3}{5}$ (3) $\frac{21}{50}$ (f) $\frac{7}{20}$

(g) $\frac{1}{25}$ (h) $\frac{21}{25}$ (i) $\frac{2}{3}$ (j) $\frac{7}{40}$ (k) $\frac{9}{80}$ (l) $\frac{79}{300}$

Page 51
(a) 50%   (b) 40%   (c) 25%   (d) 70%   (e) 83.33%
(f) 26.67%   (g) 96%   (h) 6%   (i) 150%   (j) 52.12%
(k) 140.44%   (l) 365.58%

Page 52
1. 16.67%   2. 73.33%   3. Unskilled = 71.11%, skilled = 13.33%, clerical = 15.56%   4. 83.33%

Page 52
(a) 3   (b) 17.68   (c) 21   (d) 187.05

Page 54–55
1. (a) £1.20   (b) 14 g   (c) 50p   (d) 190   (e) 0.396 lb
(f) 2.4 m
2. (a) £4.25   (b) £376.25   (c) 50 g   (d) 3 km   (e) 28.2
(f) £13 750

3. (a) £120 (b) £663.16 (c) £6.67 (d) 29.41 g
(e) 12.90 cm (f) 38.89 ml
4. 43% 5. 16% 6. (a) 75% (b) 187.50 g 7. £174.78
8. (a) 91p (b) 83p 9. £744.19 10. 840 g 11. 10% 12. 25
eggs 13. £5467.50 14. (a) £50.58 (b) £14.88 (c) £32.26
15. (a) £63.54 (b) £110.70 (c) £152.15 16. £1040
17. (a) £67.31 (b) £46.81

Measuring lengths, areas and volumes

Page 57–58
1. (a) 17 m (b) 26.02 cm (c) 11.95 cm (d) 26.9 m
(e) 13.32 cm
2. (a) 84 roses (b) £189.00 (c) 0.7 m

Page 62
A = 21 sq m, B = 111.78 sq cm, C = 1000 sq cm

Page 64
1. (a) 4.56 sq m (b) 26 sq m (c) 8.45 sq cm
2. (a) 20 sq cm (b) 29.08 sq m (c) 27.50 sq m

Page 67
(a) 9.43 cm (b) 22 m (c) 16.34 cm (d) 47.22 cm
(e) 84.02 cm

Page 68
(a) 50.27 sq cm (b) 149.59 sq m (c) 346.41 sq cm
(d) 38 018.2 sq cm (e) 514.79 sq m

Tables and charts

Page 70
1. £7.00 2. £96.00

Page 72
(a) 3.38 pm (b) 10.05 pm (c) 12.12 pm (d) Midnight or
12.00 am (e) 3.21 am

Page 73

(a) 09.32   (b) 12.02   (c) 15.45   (d) 23.59   (e) 00.30

Page 76

1. Underweight
2. 8–11 stone

Answers to Section Two

Cooking

Page 80–81

Potato salad

1 lb potatoes
8 tbspoon mayonnaise
1 tbspoon lemon juice
½ tspoon salt
2 tbspoon chives
4 tbspoon chopped leeks

Chicken pimento

15 g butter
200 g can pimentos
110 g cheese
150 ml double cream
150 ml milk

Page 82

Cheese and mushroom pie

	2 people	15 people
pastry	3.00 oz	22.5 oz
butter	0.75 oz	5.5 oz
cornflour	0.40 oz	3.0 oz
curry powder	½ tspoon	3.75 tspoon
milk	0.25 pints	2.0 pints
cheese	2.00 oz	15.0 oz
mushrooms	1.3 oz	10.0 oz
tomatoes	1	7

Page 83

1.	30 biscuits	50 biscuits
flour	375 g	625 g
sugar	125 g	210 g
margarine	250 g	420 g

2. 11.25 lb

3. *Chicken casserole*

	18 people	25 people
onions	9	12.5
celery	13 sticks	19 sticks
mushrooms	1 lb	1.5 lb
bacon	$\frac{1}{2}$ lb	12.5 oz
oil	$4\frac{1}{2}$ tbspoon	6 tbspoon
butter	4.5 oz	6.25 oz
chicken	18	25
flour	13.5 tbspoon	19 tbspoon
stock	3.25 pints	4.75 pints
tomatoes	4 × 15 oz cans	6 × 15 oz cans

Page 84–85

1. 10.41 am   2. 5.57 pm

3. (i) 2 hrs   (ii) 2 hrs 30 mins   (iii) 1 hr 45 mins   (iv) 2 hrs 15 mins   (v) 1 hr 49 mins   (vi) 2 hrs 4 mins   (vii) 2 hrs 53 mins

Page 85

19%

Page 86

(i) 6 kg   (ii) 4.92 kg   (iii) 4.2 kg   (iv) 210 g

Shopping

Page 88

110 ml in column D

Page 90

(a) £4.26   (b) £6.79   (c) £34.26   (d) £226.29
(a) £4.57   (b) £7.29   (c) £36.77   (d) £242.90

Page 91

(a) £2.80   (b) £7.22   (c) £16.80   (d) £27.54   (e) £77.06

Motoring

Page 93
£212.28

Page 95
£406.24

Page 96
Total = £3168.29, average £0.13

Page 99

Year	Depn. £	Value £
1	937	4683
2	781	3902
3	650	3252
4	542	2710
5	452	2258
6	376	1882

Page 101
1. £295
2. £127

Travelling

Page 104
(a) 365 mls   (b) 153 mls   (c) 327 mls   (d) 116 mls

Page 104
473 miles

Page 106
32.07 mpg

Page 107
10.72 gallons, £19.62

Page 108
1. Chepstow to Hereford 29.12 mph
   Hereford to Shrewsbury 33.13 mph
   Shrewsbury to Liverpool 36.10 mph

2. Sheffield to Manchester 43.53 mph
   Manchester to Preston 71.54 mph
   Preston to Blackpool 26.84 mph

Page 109
(a) 1 hr 43 mins   (b) 1 hr 49 mins

Page 110
(a) 9 miles   (b) 30.75 miles   (c) 446.25 miles

Page 111
Milan to Venice 173 miles
Venice to Rome 334 miles
Rome to Milan 356 miles

Page 111
29.87 mpg

Page 112
(a) £18.43   (b) £71.04   (c) £495.84

Working

Page 114
1. £9838.40   2. £713.42   3. £625.00

Page 115
1. £189.15   2. £2.85   3. 23 hours

Page 116
(a) £2.95   (b) £6.87   (c) £11.97   (d) £27.45

Page 116
£9485

Page 117
(a) £1722.50   (b) £2808.75   (c) £5005   (d) £7233

Page 118
1. £108   2. £255.45

Page 119–120
1. (a) £84.50   (b) £201.00   (c) £114.00   (d) £49.00
2. £702.63

Page 120
£11 053.12

Page 121–122
1. Pension = £4743.90, Lump sum = £14 231.72
2. Pension = £2834.22, Lump sum = £8502.66

Gardening

Page 124
20 sq m

Page 125

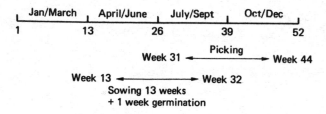

Page 126
72.36 sq m

Page 127
41.9 g

DIY

Page 129
1. 4 panels, length 820 cm
2. 3 panels, length 489 cm
3. 6 panels, length 1650 cm

Page 130
1. 36 bricks a course. Order 216
2. 55 bricks a course. Order 1460

Page 131–132
20 sq m costs £232.65

Page 133
1. (3.69) 4 × 1 litres.
2. (7.78) 8 × 10 litres

Page 134
1. 11 rolls.   2. 6 rolls.   3. 10/11 rolls.

Page 136
1. kg bags – cement 10 sand 20 aggregate 30
2. kg bags – cement 20 sand 50 aggregate 70
3. kg bags – cement 5 sand 10 aggregate 15

Housing

Page 138
(a) £1680   (b) £3818.75   (c) £3398.63

Page 138
(a) £1911   (b) 3047.50   (c) 6676.25

Page 140
Selling cost = £2805, Buying cost = £4440

Borrowing

Page 142
1. £9520   2. £6664

Page 144
Loan 1 = 49.79%, Loan 2 = 45.67%

Page 145–146
1. £254.92   2. £193.88

Banking

Page 149
£1.24

Page 151
Interest due = £24.53, Balance = £668.18

Page 154
Balance = − £24.11

Budgeting

Page 157
1. Rent £1735, Community charge £297, Insurance £328
2. +£10

Page 158
Electricity £194, Telephone £134, Food £857, Car £661

Page 159

	Monthly £	Annual £
Electricity	24.74	297
Telephone	13.05	157
Gas	18.33	220

Page 161
1. £33   2. £31   3. £34   4. £29   5. £37

Page 163
(i) (a) −£38.10   (b) +£21.72   (c) +£21.32
(ii) £204   (iii) 41%

Appendix

Table which shows the multiplication of numbers 1 to 10

	1	2	3	4	5	6	7	8	9	10
1	1	2	3	4	5	6	7	8	9	10
2	2	4	6	8	10	12	14	16	18	20
3	3	6	9	12	15	18	21	24	27	30
4	4	8	12	16	20	24	28	32	36	40
5	5	10	15	20	25	30	35	40	45	50
6	6	12	18	24	30	36	42	48	54	60
7	7	14	21	28	35	42	49	56	63	70
8	8	16	24	32	40	48	56	64	72	80
9	9	18	27	36	45	54	63	72	81	90
10	10	20	30	40	50	60	70	80	90	100

Conversion formulae

To Convert	Multiply By
Inches to C'metres	2.5400
C'metres to Inches	0.3937
Feet to Metres	0.3048
Metres to Feet	3.2810
Yards to Metres	0.9144
Metres to Yards	1.0940
Miles to Kilometres	1.6090
Kilometres to Miles	0.6214
Cu inches to Litres	0.0613
Litres to Cu inches	61.0300
Gallons to Litres	4.5460
Litres to Gallons	0.2200

Ounces to Grams	28.3500
Grams to Ounces	0.0353
Pounds to Grams	453.6000
Grams to Pounds	0.0022
Pounds to Kilograms	0.4536
Kilograms to Pounds	2.2050
Tons to Kilograms	1016.000
Kilograms to Tons	0.0009

Temperature

°F	32	40	50	60	70	75	85	95	105	140	175	212
°C	0	5	10	15	20	25	30	35	40	60	80	100

Speeds

mph	20	30	40	50	60	70	80	90	100
km/h	32	48	64	80	96	112	128	144	160

Temperature Conversion Chart

Celsius	Fahrenheit	Gas Mark
110	225	$\frac{1}{4}$
130	250	$\frac{1}{2}$
140	275	1
150	300	2
170	325	3
180	350	4
190	375	5
200	400	6
220	425	7
230	450	8
240	475	9